Lúcia H. G. Chaves
Ana Carolina F. Vasconcelos

Castor bean

Lúcia H. G. Chaves
Ana Carolina F. Vasconcelos

Castor bean

Collection of agronomic research in the semi-arid northeast of Brazil

ScienciaScripts

Cover image: www.ingimage.com

This book is a translation from the original published under ISBN 978-3-330-75908-4.

Publisher:
Sciencia Scripts
is a trademark of
Dodo Books Indian Ocean Ltd. and OmniScriptum S.R.L publishing group

120 High Road, East Finchley, London, N2 9ED, United Kingdom
Str. Armeneasca 28/1, office 1, Chisinau MD-2012, Republic of Moldova, Europe
Printed at: see last page
ISBN: 978-620-8-30667-0

INDEX

CHAPTER 1

INTRODUCTION

The castor bean tree *(Ricinus communis* L.), which probably originated in Asia and was brought to Brazil by the Portuguese during colonization, is important because of the many applications it has in the modern world. Its oil, for example, is the basis for a wide range of industrial products, and the cake, the product of oil extraction, is used as a fertilizer and soil conditioner and, if made non-toxic, serves as a protein source for animal feed.

In the period 1980/1985, Brazil had the second largest area under castor bean cultivation in the world (26% of the world area) and was the world's second largest producer in terms of quantity produced, accounting for 26% of world production (Santos et al., 2001). However, from 1986 onwards, it lost position, arriving in 1999 with only 8% of the world's cultivated area and responsible for only 2% of the amount of castor beans produced in the world.

Brazil's loss of competitiveness in the world castor bean market is explained, according to Savy Filho et al. (1999), by the inability of Brazilian farmers to use better technology, expressed in terms of the use of industrial inputs (such as fertilizer), improved seeds or even better soil preparation, planting and harvesting systems.

The great demand for energy resources by production and transportation systems, coupled with the scarcity of fossil fuels, has motivated the development of technologies that make it possible to use renewable energy sources. Among the possible alternative fuels that can be obtained from renewable biomass, which are potentially capable of running internal combustion engines with compression ignition, the most viable is Biodiesel. Biodiesel is also seen as a viable alternative for mitigating the CO_2 emissions characteristic of traditional fossil energy sources.

On December 6, 2004, based on the conclusions about the potential of biodiesel to contribute favorably to solving fundamental social, economic, environmental, strategic and even geopolitical issues for the country, the Federal Government, through the National Program for the Production and Use of Biodiesel (PNPB), decided to introduce biodiesel into the Brazilian energy matrix.

Castor bean was one of the crops chosen by the Federal Biodiesel Program to provide the raw material for its production, a biofuel that is seen as renewable and less polluting than its fossil fuel competitor, diesel. From then until recently, the Brazilian government encouraged the planting of biodiesel, especially in poor regions of Brazil, signaling that this should be the main oilseed in the still timid process of replacing Brazilian diesel. In response to the launch of various programs at different levels of government, Brazil has managed to achieve a major recovery in national castor bean production compared to the harvests of the last ten years.

In the state of Bahia alone, the main producer of castor beans, the 2005 crop of 169,800 tons was 48.79% more than the 2004 crop. The area planted increased by 14.6%, from 147.7 thousand hectares in 2004 to 169.3 thousand hectares in 2005. Productivity rose 29.84% from 773 kg/ha to 1003 kg/ha, the highest yield in the last 14 years. In Brazil that same year, according to the IBGE, the area planted with castor beans reached approximately 250,000 hectares. However, according to IBGE data, the area planted decreased in 2015, reaching approximately 85,000 hectares.

However, due to various factors, including the inadequate rainfall in recent years, the reduction in cultivation areas, the difficulty of marketing the product, the inadequate use of agricultural inputs, with little or no use of fertilizers (Azevedo et al... 1997 and 2001), among others, production has decreased, 1997 and 2001), among others, have reduced all production indices, i.e. in 2014/2015, in the state of Bahia, the planted area was 70.1 hectares, with production of 44.9 thousand tons and productivity of 640 kg/ha (Source: 4th Survey of the 2015/16 harvest / Conab). Similarly, funding for research from development agencies has decreased.

In addition, the castor bean production chain is still in its infancy; productivity is low in the main producing region (the Northeast); the scarcity and cost of labor make the harvesting process very expensive; its oil, in addition to not being edible and having a higher production cost, has a high density and viscosity, disfavoring the production of biodiesel in favor of the chemical industry due to its high lubricating power (Silva et al., 2015).

However, Brazil is in a position to increase the supply of biodiesel from castor beans, especially as a development tool for the semi-arid areas of the Northeast region, where the crop is well adapted, with abundant labor, favorable climate and soils, and low production costs. However, national castor bean production needs to be increased.

In this sense, with the encouragement of the Brazilian government, a lot of agronomic research has been carried out, the results of which show that castor bean cultivation presents favorable results in many respects and, therefore, should not be forgotten. For this reason, the aim of this paper is to summarize, as a case study, a collection of our research carried out between 2009 and 2012.

The case studies are separated into chapters according to the topic covered in each one: Chapter 1 introduces the topic covered in this book; Chapter 2 presents the case studies on mineral fertilization of papaya; Chapter 3 deals with nitrogen fertilization versus water consumption; Chapter 4 presents the use of organic fertilization on castor bean trees; Chapter 5 discusses the phytoremediation of contaminated soils using castor beans; and finally, Chapter 6 presents the final considerations on the results of the case studies presented in this book.

CHAPTER 2

MINERAL FERTILIZATION FOR PAPAYA

The papaya plant is demanding of nutrients, especially nitrogen, potassium, phosphorus, calcium and magnesium. For a yield of 1000 kg/ha of seeds, 64 to 123 kg of nitrogen (N), 14 to 44 kg of phosphorus (P_2O_5) and 52 to 160 kg of potassium (K_2O) are removed from the soil. For this reason, it must be grown in soils with good natural fertility or with a supply of organic or mineral fertilizers in order to produce well. A crop of the Campinas cultivar, which produced 2,859 kg/ha in Sao Paulo, extracted 255 kg/ha of nitrogen, 34 kg/ha of phosphorus (P_2O_5), 211 kg/ha of potassium (K_2O), 39 kg/ha of calcium (CaO) and 45 kg/ha of magnesium (MgO). In this example, the amount of nutrients extracted is very high, which is why productivity can be greatly reduced if the nutrients exported are not replaced by adequate fertilization in the following harvest (Severino et al., 2006).

There are several cultivars and hybrids of papaya available for planting in our country, which differ in certain characteristics and nutritional requirements. However, until recently, the appropriate macronutrient fertilization of this crop in different planting regions was still little known, which is why, since the beginning of the PNPB, there has been an incentive for research, especially in relation to nitrogen (N), phosphorus (P) and potassium (K) fertilization and crop management. Based on the above, several studies have been carried out looking at the behaviour of the castor bean cultivars BRS-188 Paraguaçu and BRS-149 Nordestina, which have been recommended for the Northeast region of Brazil, in relation to mineral fertilization (Case Studies 1 and 2).

When it comes to micronutrients, such as zinc and copper, the situation is even more aggravating since, although some studies have already been carried out on these nutrients (Nakagawa et al., 1986; Paulo et al., 1989; Souza and Natale, 1997; Lange et al., 2005; Severino et al., 2006), the doses of these elements have not yet been adjusted for castor bean cultivation. For this reason, an experiment was set up to evaluate the effect of increasing doses of zinc and copper on the development of castor bean plants and then, based on the results of this experiment, another was set up combining the best results of zinc and copper with the best dose of potassium from previous experiments (Case Studies 4 and 5).

CASE STUDY 1

This case study shows research aimed at evaluating the development of the BRS 188 Paraguaçu and BRS 149 Nordestina cultivars in relation to increasing doses of Nitrogen (N), Phosphorus (P) and Potassium (K).

Three experiments were set up in a greenhouse belonging to the Department of Agricultural Engineering at the Federal University of Campina Grande, using plastic pots with a capacity of 100

L filled with 80 kg of a typical Eutrophic Regolithic Neosol with the following characteristics: pH (H_2O) = 6.4; Ca = 2.41 cmolc kg^{-1} ; Mg = 2.37 cmolc kg^{-1} ; Na = 0.04 cmolc kg^{-1} ; K = 0.02 cmolc kg^{-1} ; H = 0.95 cmolc kg^{-1} ; Al = 0.20 cmolc kg^{-1} ; organic matter = 6.5 g kg^{-1} ; P = 21.7 mg kg^{-1} ; sand = 770.5 g kg^{-1} ; silt = 84.6 g kg^{-1} ; clay = 144.9 g kg .$^{-1}$

The experiments were set up in a completely randomized design with three replications, each with five levels, making a total of forty-five experimental units, which consisted of the application of N (40; 80; 120; 160 and 200 kg ha^{-1}), P2O5 (30; 60; 90; 120 and 150 kg ha^{-1}) and K_2O (30; 60; 90; 120 and 150 kg ha^{-1}). The increasing doses were based on the lowest recommendation for papaya made by Azevedo et al. (1997), which corresponds to 40:30:30 kg ha^{-1} of N:P_2O_5:K_2O. In the experimental units that received the doses corresponding to N, 90 and 60 kg ha^{-1} of P2O5 and K_2O were applied; in those that received the doses of P, 40 and 60 kg ha^{-1} of N and K_2O, and finally, in the others that received the doses of K, 40 and 90 kg ha^{-1} of N and P_2O_5. All the P2O5 and 30% of the recommended doses of K_2O for each experimental unit were applied in the foundation; the rest of the K and N doses were applied every two weeks until 120 days after sowing (DAS).

Each pot received five seeds of castor bean cultivar BRS188 Paraguaçu, with one plant remaining per pot after thinning. Throughout the experimental period (140 days), the soil was kept at a humidity close to field capacity, which was monitored daily using a Diviner 2000 probe through an access tube installed in each experimental unit. This experimental arrangement was also set up for the BRS 149 Nordestina cultivar. At 40, 60, 80, 100 and 120 DAS, biological parameters indicative of plant development were assessed, such as plant height and stem diameter. The evaluation of the number and length of leaves began at 40 DAS. The leaf area (LA) was calculated according to Wendt's method (1967), using the formula $\log(Y) = \Sigma\{-0.346 + [2.152 \times \log(X)]\}$, where Y is the leaf area in cm^2 and X is the length of the central vein of the leaf in cm. At the end of the experimental period, the ripe bunches were harvested and the seeds were weighed.

The data was analyzed using analysis of variance (ANOVA) and regression analysis for significant data, according to Ferreira (2011).

The plants of the BRS 188 Paraguaçu and BRS 149 Nordestina cvs submitted to treatments with N, P and K, regardless of the doses of these elements they received, showed higher growth rates up to 80 DAS, as can be seen from the slope of the curves (Figure 1); from this period onwards these rates decreased and/or stabilized, i.e. growth occurred more slowly after 100 DAS. It is worth noting that the time when the decrease in height growth rate begins is the same time when the first inflorescences start to appear. The halt in vegetative growth, due to the acceleration of productive growth, occurs because the photoassimilates produced are directed towards the productive organs; the forming fruit act as photoassimilate drains (Taiz and Zeiger 2004).

At the end of the crop cycle, the heights of BRS 149 Nordestina plants subjected to increasing doses of N, P and K varied from 67.0 (40 kg N ha^{-1}) to 72.3 cm (200 kg N ha^{-1}); from 61.0 (30 kg P2O5 ha^{-1}) to 70.0 cm (150 kg P2O5 ha^{-1}) and from 67.3 (30 kg K2O ha^{-1}) to 70.3 cm (150 kg K2O ha^{-1}), respectively. In the BRS 188 Paraguaçu plants, the heights varied on average from 71.4 (40 kg N ha^{-1}) to 82.1 cm (200 kg N ha^{-1}); from 59.0 (30 kg P2O5 ha^{-1}) to 67.5 cm (150 kg P2O5 ha^{-1}) and from 65.4 (30 kg K2O ha^{-1}) to 68.3 cm (150 kg K2O ha^{-1}). In general, it can be seen that the treatments with P affected castor bean growth more than those with K. The lack of these elements and/or the deficiency of N and K in the P treatments and the deficiency of N and P in the K treatments reduce their growth, probably by reducing nutrient absorption, photosynthetic activity and the translocation of carbohydrates (Taiz and Zeiger 2004).

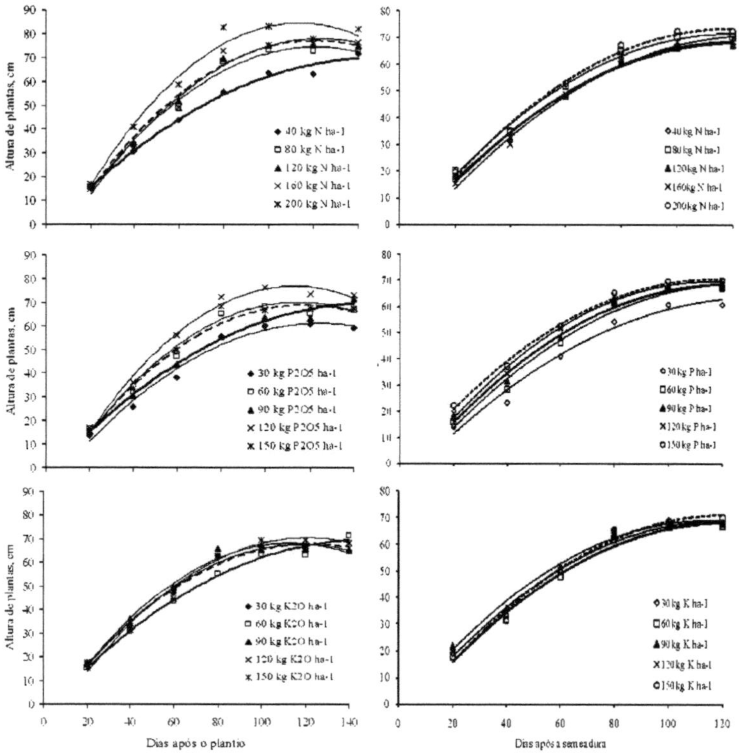

Figura 1. Height of plants subjected to different treatments with N, P2O5 and K2O, as a function of time. Thicker curves correspond to the basic treatment 40:90:60 kg ha-^{1} of NT2()<K2O. Dotted curves correspond to the highest dose of each nutrient studied. The column on the left corresponds to the BRS-188 Paraguaçu data; on the right corresponds to the BRS-149 Nordestina data.

According to the analysis of variance (Table 1), the N treatments only had a significant effect at 5%

probability on the height of the BRS 188 Paraguaçu plants at 80 DAS and, according to the regression analysis, the trend was linear (y = 0.1463N + 52.289) with a coefficient of determination (R^2) of 0.90.

Table 1. Summary of the analysis of variance for plant height, stem diameter and leaf area at 40, 60, 80, 100 and 120 days after sowing (DAS) of the BRS 149 Nordestina and BRS 188 Paraguaçu cultivars, subjected to nitrogen, phosphorus and potassium treatments.

	Days after sowing									
Treatm	40		60		80		100		120	
ent	BRS 149	BRS 188	BRS 149	BRS 188	BRS 149	BRS 188	BRS 149	BRS 188	BRS 149	BRS 188
				Plant height						
N	ns	ns	ns	ns	ns	*	ns	ns	ns	ns
P	*	ns	ns	*	ns	ns	*	ns	*	ns
K	ns	ns	ns	ns	ns	ns	ns	ns	ns	ns
				Stem diameter						
N	ns	ns	ns	*	*	*	*	ns	*	*
P	ns	ns	ns	ns	**	ns	ns	ns	ns	ns
K	ns	ns	ns	ns	ns	ns	ns	ns	ns	ns
				Leaf area						
N	ns	*	*	*	ns	ns	**	ns	*	*
P	ns	ns	ns	ns	ns	ns	*	ns	ns	ns
K	ns	ns	ns	ns	ns	ns	ns	ns	ns	ns

*, ** significant at 0.05 and 0.01 probability, ns- not significant. Sources: Araujo et al.2009; Ribeiro et al., 2009.

The effect of phosphorus on the growth of BRS 149 Nordestina plants was significant at 40 (y = 0.0656P + 61.031), 100 and 120 DAS (y = 0.1133P + 20.764), at the 5% probability level. However, the growth of BRS 188 Paraguaçu plants was only significant at 60 DAS (y = 0.1078P + 37.172) (Table 1).

With regard to potassium, there was no significant effect of the doses used on the plant height of the two cultivars (Table 1).

Stem diameter is an important characteristic in non-destructive growth analysis and, as with plant height, its growth rate was reduced after 80 DAS (Figure 2). In the same way as height, the stem diameter of the plants that were fertilized with 40:90:60 kg ha^{-1} of N:P2O5:K2O (thicker curves) behaved in the same way over time.

In general, the average stem diameter values observed ranged from 0.5 to 2.6 cm, regardless of the treatments.

At 80 DAS, the stem diameter of cv BRS 188 Paraguaçu increased by 50.3% as a result of increasing the doses of N from 40 to 200 kg ha^{-1} ; 25.5% as a result of increasing the doses of P2O5 from 30 to 120 kg ha^{-1} and 14.8% as a result of increasing the doses of K2O from 60 to 120 kg ha^{-1} , since these doses were responsible for the smallest and largest stem diameters of the plants at this time, respectively. The stem diameter of BRS 145 Nordestina at 120 DAS increased by 12.83%, 4.29% and 3.62% corresponding to the above-mentioned doses of N, P and K, respectively. The effect of N on BRS 188 Paraguaçu was significant on stem diameter at 60, 80 and 140 DAS (Table 1). According to the regression analysis, the data for these periods showed a linear trend with coefficients of determination (R^2) equal to 0.86, 0.91 and 0.98, respectively, significant at 1% (60 DAS) and 5% probability (80 and 140 DAS) (Figure 2).

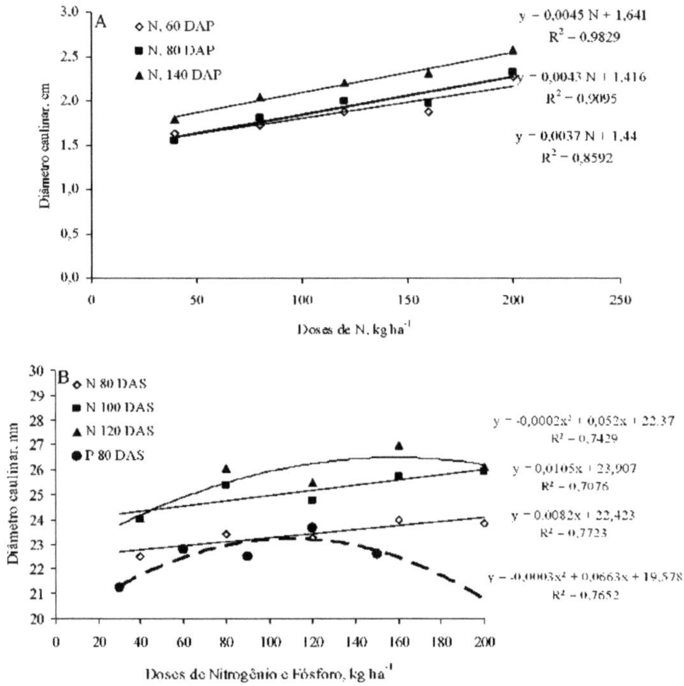

Figura 2. Stem diameter of cv BRS 188 Paraguaçu (A) as a function of increasing doses of N at 60, 80 and 140 DAS and of cv BRS 149 Nordestina (B) as a function of increasing doses of N at 80, 100 and 120 DAS and of P at 80 DAS.

Although the treatments with Pe K contributed to an increase in plant stem diameter, they had no significant effect throughout the crop cycle (Table 1). In the case of cv BRS 145 Nordestina, the stem diameter data at 80, 100 and 120 DAS, as a function of the N treatments, showed a linear trend with coefficients of determination (R^2) equal to 0.77; 0.71 and 0.74, respectively; the data as a function of the P treatments, at 80 DAS, showed a quadratic behavior, with a coefficient of determination (R2) equal to 0.76 (Figure 2).

Leaf area is one of the most important indices of plant growth, as it reflects the size of the assimilatory apparatus, which is directly related to the physiological processes of plants, especially photosynthetic activity (Oliveira et al., 2009). Measuring the leaf area of the castor bean plant is an important action for research into this crop, as it is the characteristic that best expresses whether or not the environmental conditions are suitable for the plant's development.

The highest growth rate in leaf area for the two cultivars, regardless of the treatments, generally occurred until 60 DAP and then decreased and even stabilized. The decline in leaf area as a function of growing time was due to the senescence of the leaves combined with the fact that the plants' assimilates at this time were being translocated to meet the needs of fruiting, causing a reduction in their leaf area (Taiz and Zeiger, 2004).

The N treatments had a significant effect on the leaf area of cv BRS 149 Nordestina at 60, 100 and 120 DAP, while for cv BRS 188 Paraguaçu it was significant at 40, 60 and 120 DAP (Table 1) and varied linearly (Figure 3). Phosphorus, unlike N, only significantly affected leaf area at 100 and 20 DAP for BRS 149 Nordestina and BRS 188 Paraguaçu, respectively (Table 1), whose values increased linearly (Figure 3). Although the leaf area increased as a result of the K treatments, especially at the beginning of the experiment, there was no significant difference between them throughout the experimental period for the two cultivars (Table 1).

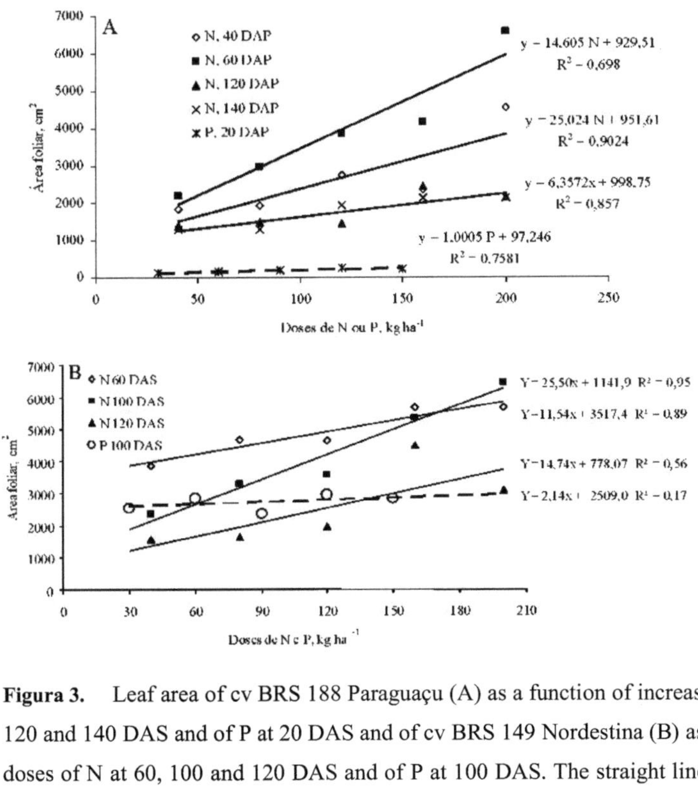

Figura 3. Leaf area of cv BRS 188 Paraguaçu (A) as a function of increasing doses of N at 40, 60, 120 and 140 DAS and of P at 20 DAS and of cv BRS 149 Nordestina (B) as a function of increasing doses of N at 60, 100 and 120 DAS and of P at 100 DAS. The straight line corresponding to the N

data at 120 DAS for BRS 188 Paraguaçu is not shown in the figure because it is superimposed on the one corresponding to 140 DAS.

The leaf area values of the BRS 188 Paraguaçu plants that received N treatments at 60 DAS ranged from 2,197 cm^2 (40 kg N ha^{-1}) to 6,592 cm^2 (200 kg N ha^{-1}), i.e. there was an increase of around 200 % in leaf area as a result of the increase in N doses from 40 to 200 kg ha^{-1} . The leaf area of the plants that received the P treatments ranged from 1,647 cm^2 (30 kg P2O5 ha^{-1}) to 2,556 cm^2 (150 kg P2O5 ha^{-1}). However, the 120 kg P2O5 ha^{-1} treatment provided the greatest leaf area, i.e. 3,113 cm^2 , which was an 89% increase compared to the lowest dose used. With increasing doses of K, the leaf area varied from 2,617 cm^2 (30 kg K2O ha^{-1}) to 2,951 cm^2 (150 kg K2O ha^{-1}), which corresponded to an increase of 13%. However, at 60 DAS, the plants treated with 120 kg K2O ha^{-1} had the lowest leaf area, 2029.67 cm^2 . Comparing this value with the highest observed, it can be said that in this case there was an increase in leaf area of 45 %, which indicates that K is the element that least influences the increase in leaf area.

In the case of cv BRS 149 Nordestina, at 80 DAS the leaf area values of the plants submitted to increasing doses of N, P and K varied from 3846.80 (40 kg ha^{-1}) to 5673.81 cm^2 (160 kg ha^{-1}) (47.49 % increase); 3804.45 (120 kg ha^{-1}) to 4361.11 cm^2 (30 kg ha^{-1}) (14.63 % increase) and from 3711.59 (90 kg ha^{-1}) to 3846.80 cm^2 (60 kg ha^{-1}) (3.64 % increase), respectively. These values are higher than those found for cv BRS 188 Paraguaçu mentioned above.

Fertilization affected not only the development of cv BRS 188 Paraguaçu, but also seed production (Figure 4A), which was statistically significant, at the 10% probability level, only as a function of increasing doses of N.

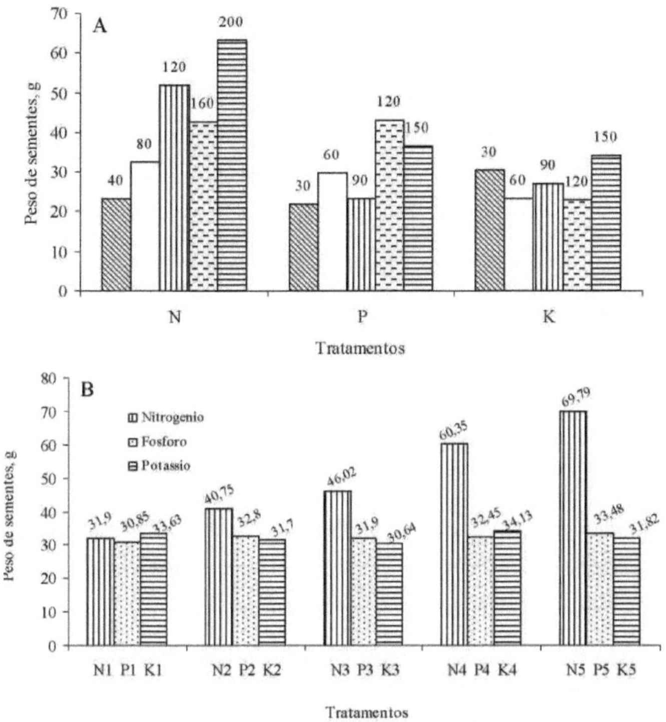

Figura 4. Seed yield of cv BRS 188 Paraguaçu (A) and cv BRS 149 Nordestina (B) as a function of increasing doses of N, P2O5 and K2O. The values above the columns correspond to the doses of each element in kg ha^{-1} The seed yield of the plants that received 200 kg N ha-1 (63.23 g) was 173.01 % higher than the yield of the plants that received 40 kg N ha^{-1} (23.16 g). Among the doses of P used, the one corresponding to 120 kg P2O5 ha^{-1} was the one that gave the highest seed yield (43.07 g), however, the increase in this yield compared to the lowest dose of the element used (21.90 g) was 96.67 %. Considering the doses of K, the increase in seed yield provided by the highest dose (34.21 g) compared to the lowest (30.29 g) was 12.94 %, with lower yields occurring at intermediate doses. As can be seen in Figure 4B, the seed weights of the phosphorus and potassium treatments hardly varied; those of the nitrogen treatments, which were significant at 1% probability, increased linearly as a function of the increasing doses of the element (Figure 5), reaching a value of 69.79 g with the highest dose of nitrogen, corresponding to an increase of 118.80% in relation to the lowest dose of the element (31.90 g). These results corroborate the seed production of the BRS 188 Paraguaçù cv.

Fertilization affected not only the development of cv BRS 188 Paraguaçu, but also seed production (Figure 4A), which was statistically significant, at the 10% probability level, only as a function of increasing doses of N.

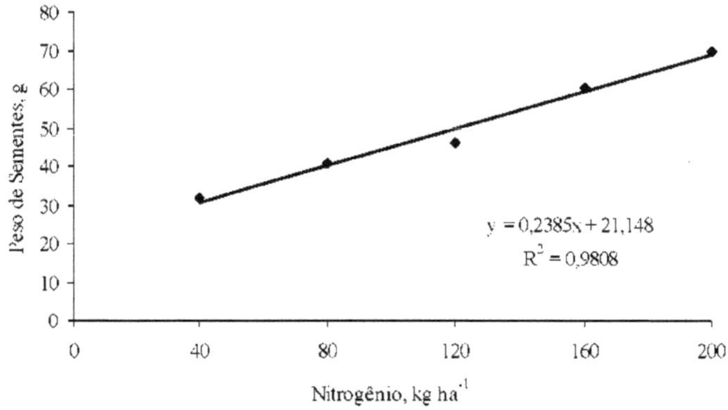

Figura 5. Variation in seed weight of cv BRS 149 Nordestina as a function of increasing doses of nitrogen. Sources: Chaves and Araujo, 2011.

According to the results, unbalanced fertilization affected the development and seed production of the papaya tree. Among the three nutrients evaluated, the application of increasing doses of nitrogen was the most beneficial for the development and production of papaya seeds.

The increasing doses of nitrogen showed the following order of benefits: leaf area > seed production > stem diameter > plant height.

The best responses from the papaya tree were the doses of 200 kg ha^{-1} of N; 120 kg ha^{-1} of P2O5 and 150 kg ha^{-1} of K2O, which should be used in future studies to assess their effect when applied together.

References

Araujo, D. L.; Chaves, L. H. G.; Mesquita, E. F.; França, C. P. Growth of papaya tree cultivar BRS-149 Nordestina fertilized with nitrogen, phosphorus and potassium. Engenharia Ambiental, v.6, n. 3, p. 685-702, 2009.

Azevedo, D. M.; Lima, E. F.; Batista, F. A. S.; Beltrao, N. E. M.; Soares, J. J.; Vieira, R. M.; Moreira, J. A. M. Recomendações técnicas para o cultivo da mamoneira (Ricinus communis L.) no nordeste do Brasil. Campina Grande: Embrapa-CNPA, 1997. 52 p. (Technical Circular, 25).

Chaves, L. H. G.; Araujo, D. L. Phytomass and production of the BRS Nordestina papaya tree fertilized with NPK. Engenharia Ambiental, v. 8, n. 1, p. 222-231, 2011.

Ferreira, D. F. Sisvar: a computer statistical analysis system. Ciência e Agrotecnologia, v. 35, p.1039-1042, 2011.

Oliveira, F. A.; Oliveira Filho, A. F.; Medeiros, J. F.; Almeida Junior, A. B.; Linhares, P. C. F. Initial development of papaya under different sources and doses of organic matter. Caatinga, v. 22, n. 1, p.

206-211, 2009.

Ribeiro, S.; Chaves, L. H. G.; Guerra, H. O. C.; Gheyi, H. R.; Lacerda, R. D. Response of the papaya cultivar BRS-188 Paraguaçu to the application of nitrogen, phosphorus and potassium. Revista Ciência Agronòmica, v. 40, n. 4, p. 465-43, 2009.

Taiz, L.; Zeiger, E. Plant physiology. 3.ed. Porto Alegre: Artmed, 2004. 719p.

Wendt, C. W. Use of a relationship between leaf length and leaf area to estimate the leaf area of cotton (Gossypium hirsutum L.), castor (Ricimus communis L.) and sorghum (Sorghum vulgare L.). (1967). Agronomy Journal, v.59, n.5, p.485-487, 1967.

CASE STUDY 2

In order to evaluate the effects of increasing doses of N, P and K, two experiments were carried out in a greenhouse belonging to the Department of Agricultural Engineering at the Federal University of Campina Grande, from April to October 2008, using plastic pots with a capacity of 100 L, perforated at the base to allow for drainage. These pots were filled with 78 kg of a sandy loam soil (Neossolo Regolitico Eutròfico tipico), collected from the topsoil, whose characteristics, according to the methods recommended by Embrapa (1997): pH (H_2O) = 6.4; Ca = 2.41 cmolc kg^{-1} ; Mg = 2.37 cmolc kg^{-1} ; Na = 0.04 cmolc kg^{-1} ; K = 0.02 cmolc kg^{-1} ; H + Al = 0.95 cmolc kg^{-1} ; Al = 0.20 cmolc kg^{-1} ; sum of bases = 4.84 cmolc kg^{-1} ; CTC = 5.99 cmolc kg^{-1} ; V = 80.8%; percentage of aluminum saturation = 3.97%; MO = 6.5 g kg^{-1} ; P = 21.7 mg kg^{-1} ; sand = 770.5 g kg^{-1} ; silt = 84.6 g kg^{-1} ; and clay = 144.9 g kg^{-1} . The experiments were set up in a randomized block design (DBC) with eight treatments and three replications. The treatments consisted of a combination of two doses of N (N1=200 and N2=300 kg/ha), two of P (P1=150 and P2=250 kg/ha) and two of K (K1=150 and K2=250 kg/ ha), making a total of 24 experimental units for each experiment. The increasing doses of N, P2O5 and K2O were based on the highest recommendation for papaya made by Ribeiro et al. (2009) and Araujo et al. (2009), which corresponded to 200:150:150 kg ha^{-1} of N:P2O5:K2O. The fertilizers used were ammonium sulphate, simple superphosphate and potassium chloride. All the P2O5 and 30% of the recommended doses of K2O for each experimental unit were applied in the foundation; the remaining doses of K (70%) and N (100%) were applied in eight equal installments, fortnightly until 120 DAS. In both experiments, each pot received five seeds and, after thinning, one plant remained per pot. Throughout the experimental period, the soil was kept at field capacity moisture, which was monitored daily using Time Domain Reflectometry (TDR). The pots were irrigated using a drip system, using water supplied by the Paraiba State Water Company (CAGEPA).

At 20, 40, 60, 80, 100, 120 and 140 days after planting (DAS), the variables related to plant growth were assessed (plant height, stem diameter at the base, number and length of leaves). The leaf area

(LA) was calculated according to Wendt's method (1967), using the formula $\log (Y) = \Sigma\{-0.346 + [2.152 \times \log (x)]\}$, where Y is the leaf area in cm^2 and x is the length of the central vein of the leaf in cm.

At the end of the experimental periods (197 DAS and 203 DAS for BRS Paraguaçu and BRS Nordestina, respectively), the plant material was harvested and dried in an air circulation oven at a temperature of 60 °C until it reached a constant mass. The total dry mass (MST) was obtained by adding up the weights of each of these parts, MSF (leaves), MSC (stems), MSR (roots). Crop production was represented by the following parameters: raceme length (CR), number of fruits per raceme (QFR), raceme weight (PR), number of seeds per raceme (QSR) and weight of 100 seeds (P100S).

The data was statistically analyzed using analysis of variance (ANOVA) and the Tukey test at 5% probability to compare the means of the results, according to Ferreira (2011).

Significant effects of N and the interactions of N x P, P x K and N x P x K were only observed at 160 DAS, although plant heights were assessed from 20 to 160 DAS. According to the breakdown of the interaction between the elements, a significant effect was found for the doses of N combined with the P2K1 treatment; for the doses of P combined with N1K1; and for the doses of K combined with the N1P1 and N1P2 treatments (Table 1).

Table 1. Plant height (cm) of the BRS Nordestina and BRS Paraguaçu cultivars at 160 DAS.

Nitrogen	Potassium			
	K_1 (150 kg ha)$^{-1}$		K_2(250 kg ha)$^{-1}$	
	Phosphorus		Phosphorus	
	P_1(150 kg ha)$^{-1}$	P_2(250 kg ha)$^{-1}$	P_1(150 kg ha)$^{-1}$	P2 (250 kg ha)$^{-1}$
		BRS 149 Nordestina		
200 kg ha^{-1}	89.67a A α	75.33 b B β	79.00 a A β	84.33 a A α
300 kg ha^{-1}	82.67 a A α	88.67 a A α	86.00 a A α	88.67 a A α
		BRS 188 Paraguaçu		
200 kg ha^{-1}	97.67 a A α	83.33 a A α	80.67 a A α	87.00 a A α
300 kg ha^{-1}	85.00 a A α	89.67 a A α	96.00 a A α	97.00 a A α

Averages followed by the same lower-case letter in the column (nitrogen doses), upper-case letter in the rows (phosphorus doses) and Greek letters in the rows (potassium doses) do not differ according to Tukey's test at 5% probability. Source: Mesquita et al. (2012a).

The lowest dosage used among the treatments studied (N1P1K1) produced the greatest height, 89.67 cm, and the lowest height value, 75.33 cm, was obtained with the N1P2K1 treatment. These results were higher than the 72.3, 70.0 and 70.3 cm observed by Araùjo et al. (2009) with the BRS Nordestina cultivar, fertilized with 200:90:60, 40:150:60 and 40:90:150 kg ha^{-1} of NPK, respectively. However, small differences can be seen between these heights, even considering the greater amounts of fertilizer applied in this work.

The analysis of variance (ANOVA) of plant height for the BRS 188 Paraguaçu cultivar showed no influence from NPK fertilization.

When comparing the plant height of the two castor bean varieties, it was observed that the BRS Paraguaçu cultivar grew more than the BRS Nordestina. The greatest height of the BRS Paraguaçu cultivar (97.67 cm) was obtained with treatment N1P1K1 (200- 150-150 kg ha^{-1}), 21.07% more than the lowest height (80.67 cm) obtained with treatment N1P1K2, which corresponds to an increase of 100 kg ha^{-1} of K (Table 1).

The analysis of variance for both cultivars showed a significant effect ($p< 0.05$) only for the N dosages evaluated at 160 DAS on stem diameter and leaf area (Table 2).

Table 2. Summary of the analysis of variance and mean significance of stem diameter and leaf area at 160 DAS of the BRS Nordestina and BRS Paraguaçu cultivars.

Variation source	GL	Mean Square			
		Culm diameter		Leaf area	
		BRS Nordestina	BRS Paraguaçu	BRS Nordestina	BRS Paraguaçu
Block	2	0,30 ^{ns}	1,04 ^{ns}	474331^{ns}	80202 ^{ns}
N	1	22,23**	16,17*	9212170**	3167935*
P	1	0,03 ^{ns}	1,35 ^{ns}	41246^{ns}	2151129^{ns}
K	1	0,63 ^{ns}	5,51 ^{ns}	388^{ns}	1435 ^{ns}
N*P	1	0,09 ^{ns}	1,00 ^{ns}	1047997^{ns}	683957 ^{ns}
N*K	1	2,34 ^{ns}	2,87 ^{ns}	37768^{ns}	249879 ^{ns}
P*K	1	0,26 ^{ns}	3,01 ^{ns}	1862^{ns}	305453 ^{ns}
N*P*K	1	1,45 ^{ns}	1,26 ^{ns}	754606^{ns}	101683 ^{ns}
Waste	14	0,71 ^{ns}	1,90 ^{ns}	761086	995108
CV (%)		3,74	6,11	17,39	21,07
Nitrogen		Averages (mm)		Averages (cm^2 plant)$^{-1}$	
N1		21,7b	21,7 b	3056,25b	3586,33b
N2		23,6a	23,4 a	6974,62a	5884,13a

Averages followed by the same lowercase letter in the column do not differ by the Tukey test at 5% probability; GL - degree of freedom; CV - coefficient of variation; ns= not significant; **= at 1% probability; *= at 5% probability. Source: Mesquita et al. (2012a).

The treatments corresponding to the doses of P and K over the experimental period did not influence the stem diameter and leaf area of the plants. Evaluating the average stem diameters of the BRS Nordestina and BRS Paraguaçu cultivars (Table 2) over the course of the experiment, it can be seen that the N2 dosage outperformed the N1 dosage by 8.39% for BRS Nordestina and 7.54% for BRS Paraguaçu at 160 DAS, respectively. It can be seen that the diameter of the stem, throughout the experiment, grew rapidly from 20 to 80 DAS, corresponding to the period of growth and flowering; later, in the period corresponding to full fruiting, from 100 to 160 DS, the stem diameters grew slowly (Figure 1).

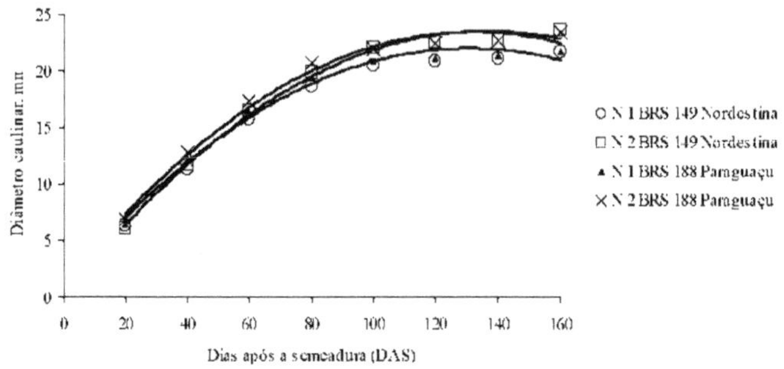

Figura 1. Culm diameter of BRS Nordestina and BRS Paraguaçu as a function of days after sowing and nitrogen levels

The effects of the N1 and N2 doses intensified over time, with accelerated growth up to 100 DAS; subsequently, from 100 to 120 DAS, for the N2 dose, there was a slower increase in this variable and then, up to 160 DAS, there was a decrease in it; for the N1 dose, after 100 DAS, there was a decrease in leaf area (Figure 2).

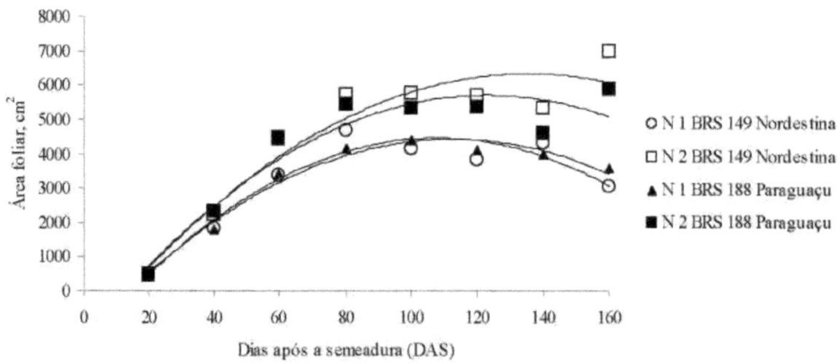

Figura 2. Leaf area of the BRS Nordestina and BRS Paraguaçu cultivars as a function of days after sowing and nitrogen doses

The decline in leaf area was probably due to leaf senescence and the fact that assimilates were being translocated by the plants at this time to satisfy fruiting needs. In general, the N2 dose (300 kg ha^{-1}) outperformed the N1 dose (200 kg ha^{-1}) at all ages studied.

The doses of N on the weight of seeds per plant (PSP), for both cultivars, had a significant effect at the 1% probability level (Table 3). The presence of P and K and the double interactions had no significant effect on yield.

Table 3. Summary of the analysis of variance for the weight of seeds per plant of the BR Nordestina and BRS Paraguaçu cultivars.

Variation source	GL	Mean Square	
		BRS 189 Nordestina	BRS 188 Paraguaçu
Block	2	698,84 ^{ns}	1254,08^{ns}
N	1	4450,19**	7103,94**
P	1	325,68 ^{ns}	73,88 ^{ns}
K	1	21,37 ^{ns}	2976,83^{ns}
N*P	1	107,65 ^{ns}	56,76 ^{ns}
N*K	1	285,17 ^{ns}	2227,4 ^{ns}
P*K	1	17,90 ^{ns}	85,01 ^{ns}
N*P*K	1	14,89 ^{ns}	42,96 ^{ns}
Waste	14	167,49	594,51
CV (%)		11,37	14,45
Nitrogen		Averages (g)	
N1		100,19 b	151,51 b
N2		127,42 a	185,92 a

Averages followed by the same lowercase letter in the column do not differ by the Tukey test at 5% probability; ns = not significant; ** = at 1% probability.

Source: Mesquita et al. (2012a).

According to what was presented, among the nutrients studied, the application of nitrogen promoted greater growth and production for the two cultivars studied, while phosphorus and potassium had no influence on the development of stem diameter, leaf area or production of the papaya cultivars. The N1P1K1 treatment (200-150-150 kg ha^{-1}) was the one that promoted the greatest height in the papaya cultivars.

The doses of N had a significant effect at the 1% probability level for all the phytomass variables studied in the two cultivars (Tables 4).

Table 4. Analyses of variance for leaf dry matter (LSM), stem dry matter (SSM), cluster dry matter (CSM), total dry matter (TSM), root dry matter (RSM) and root-to-air ratio (RAAR) of the BRS 149 Nordestina and BRS 188 Paraguaçu cultivars.

MEAN SQUARE

BRS 149 Nordestina

Source of variation	GL	MSF	MSC	MSCa	MSPA	MSR	MST
Block	2	128,00^{ns}	153,79 ^{ns}	32,32 ^{ns}	615,11 ^{ns}	31,61 ^{ns}	697 ^{ns}
Nitrogen	1	9589,20**	1122,85**	1461,09**	28783,76**	9923,48**	72508**
Phosphorus	1	46,17 ^{ns}	62,33 ^{ns}	2,84 ^{ns}	268,2^{ns}	6,22 ^{ns}	356 ns
Potassium	1	21,45 ^{ns}	50,92 ^{ns}	246,01 ^{ns}	15,34 ^{ns}	228,78^{ns}	362 ^{ns}
N*P	1	111,41 ^{ns}	457,45 ^{ns}	30,10 ^{ns}	28,57 ^{ns}	736,59^{ns}	1055 ^{ns}
N*K	1	14,36 ^{ns}	8,52 ^{ns}	876,28*	1318,53*	3,65 ^{ns}	1183 ^{ns}
P*K	1	269,00*	89,47 ^{ns}	461,65 ^{ns}	19,13 ^{ns}	693,80*	943 ns
N*P*K	1	80,55 ^{ns}	71,00 ^{ns}	21,50 ^{ns}	26,90 ^{ns}	1794,70^{ns}	2261*
Waste	14	45,70	125,61	109,61	207,65	91,59	351
CV (%)		7,55	21,55	16,75	7,06	14,92	6,99

BRS Paraguaçu

Variation source	GL	MSF	MSC	MSCa	MSPA	MSR	MST
Block	2	17,9^{ns}	12,70 ^{ns}	171,2 ^{ns}	217,65 ^{ns}	521,7 ^{ns}	98 ^{ns}
N	1	9827**	864**	1201**	26225**	3709**	49661**
P	1	0,06 ^{ns}	8,6 ns	44,2 ^{ns}	73,9 ^{ns}	15,1 ^{ns}	156^{ns}
K	1	0,8 ^{ns}	21,8 ^{ns}	9,1 ^{ns}	30,6 ^{ns}	635,4*	945 ^{ns}
N*P	1	38,2 ^{ns}	38,5^{ns}	36,5^{ns}	52,8 ^{ns}	46,4^{ns}	198 ns
N*K	1	42,7 ^{ns}	58,2 ^{ns}	140,9 ^{ns}	137,3 ^{ns}	316,6 ^{ns}	871 ^{ns}

P*K	1	10,4 ns	208,7*	14,5 ns	410,0 ns	96,1 ns	109 ns
N*P*K	1	36,3 ns	0,6 ns	236,0 ns	129,6 ns	214,2 ns	677 ns
Waste	14	58,7	38,1	123,5	314,4	73,8	300
CV (%)		8,29	17,41	11,41	8,38	16,82	6,60

ns= not significant; **= significant at the 1% probability level; *= significant at the 5% probability level. Source: Mesquita et al. (2011); Mesquita et al. (2012b).

In addition to N, the interaction between P and K had a significant effect on the leaf dry matter (DSM) of the BRS 149 Nordestina cultivar, showing that its highest content of 118.41 grams was obtained with the highest dose of N (300 kg ha^{-1}) and the lowest doses of P and K (150-150 kg ha^{-1}) (Table 5). The N2K2P2 treatment produced the highest root dry matter (RSM), 106.24 g/plant, 179.73 % more than the 37.98 g/plant obtained with the N1K2P2 treatment (Table 5), again indicating the enormous importance of nitrogen in plant growth.

Table 5 Leaf and root dry matter (g $plant^{-1}$) of the BRS Nordestina cultivar.

	Leaf dry matter			
	K1 (150 kg ha $)^{-1}$		K2 (250 kg ha $)^{-1}$	
Nitrogen	Phosphorus		Phosphorus	
	P1 (150 kg ha $)^{-1}$	P2 (250 kg ha $)^{-1}$	P1 (150 kg ha $)^{-1}$	P2 (250 kg ha $)^{-1}$
200 kg ha^{-1}	72.01 b A α	70.51 b A α	65.54 b A α	70.10 b A α
300 kg ha^{-1}	118.41 a Aα	99.14 a B α	109.53 a A α	110.98 a A β
	Root dry matter			
	K1(150 kg ha $)^{-1}$		K2(250 kg ha $)^{-1}$	
Nitrogen	Phosphorus		Phosphorus	
	P1 (150 kg ha $)^{-1}$	P2 (250 kg ha $)^{-1}$	P1 (150 kg ha $)^{-1}$	P2(250 kg ha $)^{-1}$
200 kg ha^{-1}	43.12 b A α	37.56 b A α	56.62 a A α	37.98 b B α
300 kg ha^{-1}	90.78 a A α	72.80 a B α	68.13 a B β	106.24 a A β

Averages followed by the same lower-case letter in the column, upper-case letter in the row (phosphorus at each nitrogen and potassium level) and Greek letter (potassium at each nitrogen and phosphorus level) do not differ at 5% probability using the Tukey test. Source: Mesquita et al. (2011).

The 300 kg ha^{-1} dose of N caused a 23.26 % increase in the stem dry matter (SCM) of the BRS Nordestina papaya tree compared to the 200 kg ha^{-1} dose, producing 45.12 g/plant. The treatments used, N1K2 and N2K2, produced the lowest (162.76 g) and highest (246.85 g) dry mass of the aerial part (MSPA), respectively, with a superiority of 51.66 % (Table 6).

Table 6 - Dry matter of the aerial part (g $plant^{-1}$) of the cultivar BRS 149 Nordestina, under greenhouse conditions;

Nitrogen	BRS Nordestina	
	Potassium	
	K1(150 kg ha $)^{-1}$	K2(250 kg ha $)^{-1}$
200 kg ha^{-1}	175.99 b A	162.76 b A
300 kg ha^{-1}	230.42 a B	246.85 a A

Averages followed by the same lower-case letter in the column, upper-case letter in the row and Greek letter do not differ at 5% probability using the Tukey test. Source: Mesquita et al. (2011).

Total dry matter (TMS) was significantly influenced by N and the NPK interaction. Plants fertilized with the N2P2K2 treatment had a higher average MST of 352.79 g, 79.86, 68.57, 57.93, 57.50, 17.98, 11.89 and 8.41% more than the N1P2K2 treatments; N1P2K1; N1P1K1; N1P1K2; N2P2K1; N2P1K2 and N2P1K1, corresponding to values of 196.14; 209.28; 223.37; 223.98; 299.02; 315.28 and 325.41 g/plant, respectively (Table 7).

Table 7. Total dry matter (g $plant^{-1}$) of the BRS Nordestina cultivar.

Nitrogen	BRS Nordestina			
	K_1(150 kg ha $)^{-1}$		K_2(250 kg ha $)^{-1}$	
	Phosphorus		Phosphorus	
	P_1(150 kg ha $)^{-1}$	P_2(250 kg ha $)^{-1}$	P_1(150 kg ha $)^{-1}$	P_2(250 kg ha $)^{-1}$
200 kg ha^{-1}	223.37 b A α	209.28 b A α	223.98 b A α	196.14 b A α
300 kg ha^{-1}	325.41 a A α	299.02 a A α	315.28 a B α	352.79 a A β

Averages followed by the same lower-case letter in the column, upper-case letter in the row and Greek letter do not differ at 5% probability using the Tukey test. Source: Mesquita et al. (2011).

The dry leaf mass (DSM) values of the BRS Paraguaçu cultivar ranged from 112.74 (a) to 72.27 (b) g/plant for the N2 and N1 doses (300 and 200 kg ha^{-1}), respectively, so an increase of 100 kg ha^{-1} of nitrogen provided an increase of 40.47 g/plant of dry matter, equivalent to a 55.99% increase. The doses of the inputs used, corresponding to the treatments N2P2K2 (250-250-300 kg ha^{-1}) and N1P2K1 (200-250- 150 kg ha^{-1}) gave the highest and lowest values of dry stem mass (DSM), in the order of 67.28 g/plant and 44.94 g/plant, respectively (Table 8).

Table 8. Stem dry mass (g $plant^{-1}$) of the BRS Paraguaçu cultivar under greenhouse conditions.

Nitrogen	K1 (150 kg ha $)^{-1}$		K2 (250 kg ha $)^{-1}$	
	Phosphorus		Phosphorus	
	P1 (150 kg ha $)^{-1}$	P2 (250 kg ha $)^{-1}$	P1 (150 kg ha $)^{-1}$	P2 (250 kg ha $)^{-1}$
200 kg ha^{-1}	52.50 a A a	44.94 b A a	45.07 b A a	49.96 b A a
300 kg ha^{-1}	58.52 a A a	56.69 a A a	57.98 a B a	67.28 a A a

Averages followed by the same lower-case letter in the column, upper-case letter in the row (phosphorus at each nitrogen and potassium level) and Greek letter (potassium at each nitrogen and phosphorus level) do not differ at 5% probability by Tukey's test. Source: Mesquita et al. (2012b).

The average dry mass of the aerial part (MSPA) of the BRS Paraguaçu cultivar was higher in the plants fertilized with the N2 dose (300 kg $_{ha-1}$) compared to the plants fertilized with the N1 dose (200 kg ha^{-1}), whose values were in the order of 244.78(a) and 178.67(b) g/plant, respectively. With each increase of 1 kg ha^{-1} of nitrogen, there was an increase of around 0.66 g/plant in the dry matter of the aerial part. The highest average value for root dry matter (RSM) of the BRS Paraguaçu cultivar for the interaction between nitrogen doses and potassium doses occurred in the N2K2 treatment (300 and 250 kg ha^{-1}), corresponding to 72.29 g/plant, which was 94.6, 80 and 32% higher than the N1K1, N1K2 and N2K1 treatments, corresponding to 37.13, 40.16 and 54.73 g/plant, respectively (Table 9).

Table 9. Root dry mass (g $plant^{-1}$) of the BRS Paraguaçu cultivar under greenhouse conditions.

Nitrogen	K_1(150 kg ha $)^{-1}$	K_2(250 kg ha $)^{-1}$
N_1(200 kg ha $)^{-1}$	37.13 b A	40.16 b A
N_2(300 kg ha $)^{-1}$	54.73 a B	72.29 a A

Averages followed by the same lower-case letter in the column, upper-case letter in the row and Greek letter do not differ at 5% probability using the Tukey test. Source: Mesquita et al. (2012b).

It can be seen that the highest root dry matter weights occurred independently of the nitrogen dose (N2= 300 kg/ha), so the papaya plants responded positively to nitrogen fertilization. The papaya plants fertilized with the lowest and highest doses of nitrogen (N1= 200 kg ha^{-1} ; N2= 300 kg ha^{-1}) had the lowest and highest average MST values per plant, corresponding to 217.32 (b) and 308.29 (a) g/plant, respectively. The insignificant gains in MST are probably related to the size of the pots used, which may have limited plant growth and/or the low response of the plants to the higher amount of nitrogen applied.

According to Diniz Neto et al. (2009), the castor bean plant species responds intensely to the application of fertilizers in the soil through production components and productivity. Thus, for the BRS Nordestina cultivar, there was a significant difference at the 1% probability level for the doses of nitrogen on the number of fruits per plant (NFP), seed weight per plant (PSP), number of seeds per plant (NSP), weight of 100 seeds (P100S) and bunch weight per plant (PCP) (Table 10). In the case of BRS Paraguaçu, the number of fruits per plant (NFP) and the number of seeds per plant (NSP) were influenced at the 5% probability level by the application of nitrogen doses; while the weight of seeds per plant (PSP), the weight of bunches per plant (PCP) and the length of the primary bunch (CCP) were influenced at the 1% probability level; however, there was no significant effect for the weight of 100 seeds (P100S) (Table 10).

Table 10. Mean square values, significances for Number of Fruits per Plant (NFP), Seed Weight per Plant (PSP), Number of Seeds per Plant (NSP), Weight of 100 Seeds (P100S), Bunch Weight per Plant (PCP) and Primary Bunch Length (CCP) in the BR 189 Nordestina and BRS 188 Paraguaçu castor bean trees.

Variation source	GL	Mean Square					
		BRS Nordestina					
		NFP	PSP	NSP	P100S	PCP	CCP
Block	2	10,50^{ns}	698,84 ^{ns}	239,04 ^{ns}	164,12^{ns}	597,61 ^{ns}	1,38^{ns}
Nitrogen	1	1162,04**	4450,19**	9882,04**	12,36**	6901,02**	0,66^{ns}
Phosphorus	1	18,37 ^{ns}	325,68 ^{ns}	70,04 ^{ns}	279,68^{ns}	199,46 ^{ns}	1,50^{ns}
Potassium	1	92,04 ^{ns}	21,37 ^{ns}	442,04 ^{ns}	1,33 ^{ns}	1649,87^{ns}	1,50^{ns}
N*P	1	5,04 ^{ns}	107,65 ^{ns}	3,37 ^{ns}	8,86 ^{ns}	1359,76^{ns}	2,66^{ns}
N*K	1	135,37 ^{ns}	285,17 ^{ns}	925,04 ^{ns}	569,10^{ns}	2630,27^{ns}	4,16^{ns}
P*K	1	187,04 ^{ns}	17,90 ^{ns}	1650,04^{ns}	435,96^{ns}	881,48 ^{ns}	10,66^{ns}
N*P*K	1	18,37 ^{ns}	14,89 ^{ns}	35,04 ^{ns}	0,04 ^{ns}	915,99 ^{ns}	0,16^{ns}
Waste	14	61,16	167,49	470,51	112,10	1128,07	6,33
CV (%)		15,53	11,37	14,49	13,71	17,57	11,03
Source of variation	GL -	**BRS Paraguaçu**					
		NFP	PSP	NSP	P100S	PCP	CCP
Block	2	162,50^{ns}	1254,08^{ns}	1746,12^{ns}	672,38^{ns}	940,76 ^{ns}	25,08^{ns}
Nitrogen	1	384,00*	7103,94**	3060,04*	18,06 ^{ns}	7120,47**	19,62**
Phosphorus	1	13,50 ^{ns}	73,88 ^{ns}	30,37 ^{ns}	111,62^{ns}	9,66 ^{ns}	2,22^{ns}
Potassium	1	80,66 ^{ns}	2976,83^{ns}	513,37 ^{ns}	274,86^{ns}	3536,38*	9,75^{ns}
N*P	1	24,00 ^{ns}	56,76 ^{ns}	260,04 ^{ns}	357,28^{ns}	599,70 ^{ns}	6,30^{ns}
N*K	1	60,16 ^{ns}	2227,4 ^{ns}	187,04 ^{ns}	455,70^{ns}	3472,57^{ns}	38,25*
P*K	1	16,66 ^{ns}	85,01 ^{ns}	117,04 ^{ns}	56,42 ^{ns}	276,69 ^{ns}	0,92^{ns}
N*P*K	1	0,16 ^{ns}	42,96 ^{ns}	145,04 ^{ns}	47,94 ^{ns}	13,60 ^{ns}	0,12^{ns}
Waste	14	53,45	594,51	358,60	145,17	555,41	3,41
CV (%)		16,71	14,45	14,72	9,16	13,52	12,80

GL= Degree of Freedom, ns = not significant; ** = at 1% probability level; * = at 5% probability level; MSD= Minimum Significant Difference. Source: Mesquita et al. (2011); Mesquita et al. (2012b).

The number of seeds per plant (NSP) of the BRS Nordestina cultivar was similar to the number of fruits per plant (NFP) and seed weight per plant (PSP), i.e. the plants fertilized with the highest dose of nitrogen (N2 = 300 kg ha^{-1}) had the highest number of seeds (170 seeds), an increase of 31.35 compared to the lowest dose of nitrogen (121.33 seeds). The plants fertilized with the lowest and highest doses of nitrogen (N1 = 200 kg ha^{-1} ; N2 = 300 kg ha^{-1}) had the lowest and highest average seed weight per plant, corresponding to 100.19 and 127.42 g/plant, respectively (Table 11).

Table 11. Average values (g $plant^{-1}$) for Number of Fruits per Plant (NFP), Seed Weight per Plant (PSP), Number of Seeds per Plant and Weight of 100 Seeds (P100S) and Bunch Weight per Plant (PCP) as a function of nitrogen doses.

	BRS Nordestina				
Variables	NFP	PSP	NSP	P100S	PCP
N1(200 kg/ha)	43,42 b	100,19 b	129,42 b	77,92 a	174,21 b
N2(300 kg/ha)	57,33 a	127,42 a	170,00 a	76,49 b	208,12 a
DMS	6,84	11,32	18,99	1,22	29,40

Source: Mesquita et al. (2011).

The weight of 100 seeds behaved differently from the other production components, i.e. the plants fertilized with the N1 dosage had a higher average value of 77.92 g, 1.86% more than the plants fertilized (76.49 grams) with the N2 dosage (Table 11). In any case, these values are adequate because, according to Beltrao and Azevedo (2007), medium-sized cultivars, such as BRS Nordestina, had an average of 68 g per 100 seeds. According to the above, the 300: 250: 250 kg ha^{-1} N-PK treatment was the most efficient for the dry phytomass components of the BRS Nordestina cultivar. The phosphorus and potassium treatments did not affect the phytomass of this cultivar. Due to the lack of interaction between the NPK nutrients, the highest production yield (weight of seeds per plant) was obtained with the highest dose of nitrogen and the lowest doses of phosphorus and potassium.

In addition to nitrogen, the PCP of the BRS Paraguaçu cultivar was also significantly influenced at the 5% probability level by the potassium doses (Table 10). The number of fruits per plant (NFP) obtained with the nitrogen doses, N1= 200 kg ha^{-1} and N2=300 kg ha^{-1} , were 39.75 (b) and 47.75 (a), respectively, with the former showing a superiority of 20.12% over the lower N dose. The plants fertilized with the lowest and highest doses of nitrogen (N1= 200 kg ha^{-1} ; N2= 300 kg ha^{-1}) had the lowest and highest average values for seed weight per plant (PSP), corresponding to 151.51 and 185.92 g/plant. The behaviour of the number of seeds per plant (NSP) was similar to the number of fruits per plant (NFP) and seed weight per plant (PSP), i.e. the plants fertilized with the highest dose of nitrogen (N2 = 300 kg ha^{-1}) had the highest number of seeds (139.92 seeds), an increase of 19.25% compared to the lowest dose of nitrogen (117.33 seeds). In relation to the weight of the bunch per plant (PCP), it was observed that, in addition to the significant effect of nitrogen at 1%, potassium had a significant effect at the 5% probability level. According to the split between nitrogen and potassium, the N2K2 treatment produced an average of 221.70 g/plant, 36.02, 35.84 and 27.87%

more than the N1K1, N1K2 and N2K1 treatments, corresponding to 162.98, 163.20 and 173.37 g/plant, respectively (Table 12).

Table 12. Breakdown of the nitrogen x potassium dose interaction for bunch weight (g $plant^{-1}$) in the BRS Paraguaçu cultivar.

Nitrogen	Potassium	
	K1(150 kg/ha)	K2(250 kg/ha)
N1(200 kg ha $)^{-1}$	162.98 a A	163.20 b A
N2(300 kg ha $)^{-1}$	173.37 a B	221.70 a A

Averages followed by the same lower-case letter in the column and upper-case letter in the row do not differ at 5% probability using the Tukey test. Source: Mesquita et al. (2012b).

In relation to the study of nitrogen doses with potassium dosage, there was a significant effect with K2= 250 kg ha^{-1} ; there was also a statistical difference between potassium doses and nitrogen dosage N2= 300 kg ha^{-1} for the length of the primary raceme. The following increasing order of primary raceme length was obtained: N2K2 > N1K1 > N2K1 > N1K2, with the greatest raceme length (17.25 cm) being obtained with the N2K2 treatment (Table 13).

Table 13. Breakdown of the nitrogen x potassium dose interaction for primary raceme length (cm) in the BRS Paraguaçu cultivar.

Nitrogen	Potassium	
	K1(150 kg/ha)	K2(250 kg/ha)
N1(200 kg ha $)^{-1}$	14.167 a A	12.917 b A
N2(300 kg ha $)^{-1}$	13,450 a B	17,250 a A

Averages followed by the same lower-case letter in the column, upper-case letter in the row and Greek letter do not differ at 5% probability using the Tukey test. Source: Mesquita et al. (2012b).

The treatments with the highest doses of nitrogen, phosphorus and potassium gave the highest dry phytomass values for the cultivar studied. The production yield of the BRS Paraguaçu cultivar responded more to the action of nitrogen doses than phosphorus and potassium doses. Nitrogen was the nutrient that promoted the highest yield of the cultivar studied.

References

Araujo, D. L.; Chaves, L. H. G.; Mesquita, E. F.; França, C. P. Growth of papaya tree cultivar BRS-149 Nordestina fertilized with nitrogen, phosphorus and potassium. Engenharia Ambiental, v.6, n. 3, p. 685-702, 2009.

Beltrao, N. E. M.; Azevedo, D. M. P. Physiology. In: Azevedo, D. M. P.; Beltrao, N. E. M. M. O agonegócio da mamona no Brasil. 2 ed. Campina Grande: Embrapa Algodao. Embrapa Informaçao Tecnológica, 2007, p. 140-150.

Diniz Neto, M. A.; Tàvora, F. J. A. F.; Crisòstomo, L.A.; Diniz, B. L. M. T. NPK fertilization and planting times for papaya. II - Components of the vegetative and reproductive phases. Revista Ciência Agronômica, v.40, n. 3, p. 417-426, 2009.

Embrapa. National Soil Research Center. Manual of soil analysis methods. 2.ed. Rio de Janeiro: Embrapa, CNPS, 1997. 212p.

Ferreira, D.F. Sisvar: a computer statistical analysis system. Ciência e Agrotecnologia, v. 35, p.1039-1042, 2011.

Mesquita, E. F.; Chaves, L. H. G.; Guerra, H. O. C. Phytomass and yield components of castor bean fertilized with nitrogen, phosphorus and potassium. Agrarian Journal, v. 4, n. 14, p. 344-351, 2011.

Mesquita, E. F.; Chaves, L. H. G.; Guerra, H. O. C.; Lacerda, R. D. Growth and production of two papaya cultivars under NPK fertilization. Revista Caatinga, v. 25, n. 2, p. 35-43, 2012a.

Mesquita, E. F.; Chaves, L. H. G.; Guerra, H. O. C. Phytomass and yield components of castor bean cultivar Paraguaçu fertilized with nitrogen, phosphorus and potassium. Engenharia Ambiental, v. 9, n. 2, p. 94-104, 2012b.

Ribeiro, S.; Chaves, L. H. G.; Guerra, H. O. C.; Gheyi, H. R.; Lacerda, R. D. Response of the papaya cultivar BRS-188 Paraguaçu to the application of nitrogen, phosphorus and potassium. Revista Ciência Agronòmica, v. 40, n. 4, p. 465-43, 2009.

Wendt, C. W. Use of a relationship between leaf length and leaf area to estimate the leaf area of cotton *(Gossypium hirsutum* L.), castor *(Ricimus communis* L.) and sorghum (*Sorghum vulgare* L.). (1967). Agronomy Journal, v.59, n.5, p.485-487, 1967.

CASE STUDY 3

This case study is concerned with the effect of mineral fertilization with zinc and copper on the nutritional status of the BRS 188 Paraguaçu cultivar of papaya. Two experiments were set up in a greenhouse belonging to the Department of Agricultural Engineering at the Federal University of Campina Grande, using plastic pots filled with 22 kg of Neossolo Quartzarenico. The treatments in one experiment consisted of five doses of Zn (0; 2; 4; 6 and 8 mg dm^{-3}) and in the other, five doses of Cu (0; 1; 2; 3 and 4 mg dm^{-3}) with three replications. In addition, in these experiments, the soil was fertilized with 200 kg ha^{-1} of N (ammonium sulphate), 30 kg ha^{-1} of P2O5 (triple superphosphate) and 150 kg ha^{-1} of K2O (potassium chloride). During the experimental period (130 days), one plant was kept in each pot, with humidity close to field capacity. At 90, 110 and 130 days after sowing (DAS), the fourth leaf of the main branch was removed from each plant, starting from the apex, so that the copper and zinc content could be determined. At the end of the crop, the biological parameters indicative of plant development, the dry matter of the various parts of the plant and the copper and zinc levels in the leaves and stems of the plants were assessed.

The components linked to the growth of the plants subjected to the Zn treatments, such as height and stem diameter, behaved similarly throughout the experimental period, i.e. they increased until 77 days after planting (DAP) and then practically stabilized or grew less. All the plants that did not receive zinc (control) behaved in the same way, suggesting that there was no effect from the treatments. In all treatments, the leaf area of the plants also increased until approximately 77 DAP, and then decreased. This can be considered normal, as from this time onwards the plants lost their leaves, which went into senescence, probably due to the nutritional deficiency.

Table 1 shows the average weights of the stems, leaves and bunches of plants, although the treatments had no significant effect on these parameters.

Table 1. Averages of stem, leaf and cluster data as a function of zinc doses.

Doses of Zn (mg dm)$^{-3}$	Stem (g)	Leaves (g)	Bunches (g)
0	13,33	13,41	26,09
2	11,57	13,17	24,01
4	12,47	12,39	20,17
6	11,96	12,86	24,33
8	13,16	14,56	20,90

Source: Chaves et al. (2009)

This could be explained by a number of hypotheses or even a combination of factors. One of these could be related to the doses of N, P and K that the plants received.

Despite applying amounts of N and K above those recommended by Azevedo et al. (1997), while maintaining P (30 kg ha^{-1}) within the dose recommended by the authors, a decline in the plants of all the treatments was noticed as early as 30 DAP. According to the literature (Ferreira et al., 2004), the amount of P applied was responsible for the delay in crop development, in accordance with the Law of the Minimum (Raij, 1991). Therefore, the fact that the plants did not develop properly probably meant that they did not respond to the increasing doses of zinc incorporated into the soil.

Another hypothesis would be the fact that the soil used for this study already had 10 mg dm^{-3} of Zn, which is considered a high value (Ribeiro et al., 1999) and may already be sufficient for the development of the castor bean crop under the conditions of the experiment, preventing it from responding to the treatments used. And finally, a third hypothesis would be the size of the pots used to grow the castor bean for 103 days, where the amount of soil applied to the pots was not enough to provide satisfactory development of the crop's root system and, consequently, its aerial part.

In relation to the increasing doses of copper, the height and stem diameter of the plants increased as a function of time, and these parameters would have continued to increase if the experimental period had been longer, since there was no stabilization of the values between the last two evaluation periods (77 and 103 DAP). The behavior of the leaf area of these plants was similar to that of the plants that received Zn, i.e. the leaf area increased until approximately 77 days after planting and then decreased. However, according to the analysis of variance, these treatments did not significantly influence the biological parameters indicative of plant development. This is probably because the amounts of N, P and K used in the experiment may have had an impact on plant yields. Another aspect to be taken into account is that the initial Cu content in the soil (0.2 mg dm^{-3}), which is considered low or very low, even when added to the quantities of the element supplied through the treatments, was not enough for there to be a significant response from the plants in terms of the parameters stem diameter, height and leaf area.

The significant effect of the treatments was only observed on the leaf dry weight data (5% probability), with the highest dose of copper standing out from the others (Table 2), and its behavior was best explained by the quadratic regression.

Table 2. Averages of stem, leaf and bunch data as a function of copper doses.

Doses of Zn (mg dm $)^{-3}$	Stem (g)	Leaves (g)	Bunches (g)
0	17,75	16,47	19,76
1	13,71	14,58	17,04
2	12,27	14,42	23,50
3	13,60	14,74	19,07
4	12,96	28,53	21,44

Source: Chaves et al. (2009a)

Most of the leaf zinc levels were within the range of 14 to 43 mg kg^{-1} according to the literature (Hocking, 1982; Souza and Natale, 1997). However, the copper levels in the leaves were below those found by other authors, i.e. below 7.8; 6.0; 8.8 and 8 mg kg^{-1} according to Paulo et al. (1989), Lange et al. (2005), Camargo and Zabrini (2005) and Pacheco et al. (2006), respectively (Table 3).

Table 3: Average zinc and copper levels in the fourth leaf of plants collected at 90, 110 and 130 DAP and in leaves and stems collected at the end of the experiment, as a function of zinc and copper doses.

Doses of Zn	Fourth leaf			Total leaves	Stem
	90 DAP	110 DAP	130DAP		
(mg dm $)^{-3}$	-	(mg kg $)^{-1}$ ------			
0	28,86	21,33	17.08	21,50	18,87
2	27,39	19,86	16,75	23,63	28,54
4	33,61	18,06	16,91	22,31	32,96
6	33,45	16,59	13,64	29,85	47,69
8	35,25	15,77	14,62	27,23	46,87
Leaf at 90 DAP		$Y = 0.0029 x^2 + 0.9191 x + 27.967$			$R^2 = 0.77$
Total leaves		$Y = 0.884 x + 21.37$			$R^2 = 0.63$
Stem		$Y = 3.757 x + 19.96$			$R^2 = 0.92$
Doses of Cu		Fourth leaf		Total leaves	Stem
	90 DAP	110 DAP	130DAP		
(mg dm $)^{-3}$	-	(mg kg $)^{-1}$ -------			
0	1,17	1,17	0,50	1,20	0,2
1	1,30	1,23	0,50	1,47	0,2
2	1,07	1,30	0.83	1,27	0,2
3	0,60	1,20	1,07	1,60	0,2
4	1,80	1,30	0,97	1.37	0,2
Leaf at 90 DAP			$Y = 0.156 x + 1.076$		$R^2 = 0.65$
Leaf at 130 DAP			$Y = 0.151 x + 0.472$		$R^2 = 0.82$

Source: Chaves et al. (2009b)

In relation to the time when the leaves were collected, the levels of zinc and copper decreased, probably due to a reduction in the dry matter production of the leaves. Although the plants showed no visual symptoms of zinc and copper deficiency, their development must have been influenced by some nutritional deficiency in relation to the macronutrients, especially the phosphorus content, which is responsible for the delay in development. In the case of nitrogen deficiency, the behavior of copper within plants is compromised, since this element travels both in the xylem and phloem in the form of complexes with the amino group, since the Cu ion^2 + has a strong affinity with the N atom (Marschner, 1995).

The zinc contents observed in the stem were higher than those in the leaves (Table 3), however, increasing doses of the element significantly influenced the zinc content in the leaves collected at 90 DAS (5% probability), in the total leaves (5% probability) and in the stem (1% probability). In the case of copper, there was an influence on the content of the element in the leaves collected at 90 (1% probability) and 130 (5% probability) DAS.

References

Azevedo, D. M.; Lima, E. F.; Batista, F. A. S.; Beltrao, N. E. M.; Soares, J. J.; Vieira, R. M.; Moreira, J. A. M. Recomendações tècnicas para o cultivo da mamoneira (Ricinus communis L.) no nordeste do Brasil. Campina Grande: Embrapa-CNPA, 1997. 52 p. (Technical Circular, 25).

Camargo, A. P. M.; Zabrini, A. V. Nutritional diagnosis of papaya in response to foliar fertilization in western Bahia. In: Congresso Brasileiro de Plantas Oleaginosas, Óleos, Gorduras e Biodiesel, 2, 2005. Varginha. Expanded Abstracts...Lavras: UFLA, 2005. CD Rom.

Chaves, L. H. G.; Barros Junior, G.; Lacerda, R. D.; Cabral, P. C.; Cunha, T. H. C. S. Response of papaya fertilized with zinc and copper. Engenharia Ambiental, v. 6, n. 3, p. 306-319, 2009a.

Chaves, L. H. G.; Barros Junior, G.; Lacerda, R. D.; Cabral, P. C.; Cunha, T. H. C. S. Effect of zinc and copper on the nutritional status of the papaya tree, BRS 188 Paraguaçù. Caatinga, v. 22, n. 4, p. 129-135, 2009b.

Ferreira, G. B.; Santos, A. C. M.; Xavier, R. M.; Ferreira, M. M.; Severino, L. S.; Beltrao, N. E.; Dantas, J. P.; Moraes, C. R. A. *Phosphorus* and potassium deficiency in castor bean *(Ricinus communis* L.): description and effect on crop growth and production. In: Brazilian Castor Bean Congress, 1, 2004. Campina Grande. Anais...Campina Grande: Embrapa Cotton, 2004. CD Rom.

Hocking, P. J. Accumulation and distribuition of nutrients in fruits of castor bean (*Ricinus communis* L.) Annals of Botany, v. 49, p. 51- 62, 1982.

Lange, A.; Martines, A. M.; Silva, M. A. C.; Sorreano, M. C.; Cabral, C. P.; Malavolta, E. Effect of micronutrient deficiency on the nutritional status of the papaya cultivar Iris. Pesquisa Agropecuària

Brasileira, v. 40, n. 1, p. 6167, 2005.

Pacheco, D. D.; Saturnino, H. M.; Gonçalves, N. P.; Santos, D. A.; Lopes, H. F.; Souza, R. P. D.; Dourado, I. C.; Antunes, P. D.; Ribeiro, D. P.Nutritional diagnosis for micronutrients in castor bean fertilized with NPK in plateau soil of the Rio do Jequitinhonha Basin. In: Brazilian Castor Bean Congress, 2, 2006. Aracajù. Anais...Aracaju: SAGRI, Embrapa Tabuleiros Costeiros and Embrapa Algodao, 2006. CD Rom.

Paulo, E. M. et al. Boron deficiency in castor bean. Bragantia , v. 48, p.241-247, 1989.

Marschner, H. Mineral nutrition of higher plants. 2.ed. San Diego: Academic, 1995. 902p.

Raij, B. van. Soil fertility and fertilization. Piracicaba: Agronômica Ceres, 1991. 343p.

Ribeiro, A. C.; Guimaraes, P. T. G.; Alvarez, V. V. H. Recomendaçao para uso de corretivos e fetilizantes em Minas Gerais. 5ª . Aproximaçao. Viçosa, MG, 1999. 359p.

Souza , E. C. A.; Natale, W. Effect of boron and zinc on papaya cultivation. Cientifica, v. 25, n. 2, p. 327-333, 1997.

CASE STUDY 4

Continuing the work related to the fertilization of papaya trees, and in line with the results of research into increasing doses of macronutrients (Araujo et al., 2009; Ribeiro et al., 2009; Mesquita et al., 2009) and micronutrients (Chaves et al., 2009 a, b), the aim of this study was to evaluate the growth and yield of the BRS Nordestina and BRS Paraguaçu cultivars, as a function of higher doses of potassium, zinc and copper.

To this end, two experiments were carried out in a greenhouse belonging to the Department of Agricultural Engineering at the Federal University of Campina Grande, from January to May 2009, with the papaya cultivars BRS Nordestina and BRS Paraguaçu, grown in plastic pots individually filled with 80 kg of sandy loam soil material (typical Eutrophic Regolithic Neosol), collected from the topsoil (0-20 cm deep) with the following characteristics: sand = 770.5 g kg^{-1} ; silt = 84.6 g kg^{-1} ; clay = 144.9 g kg^{-1} ; pH (H_2O) = 6.4 ; Ca = 2.41 cmolc kg^{-1} ; Mg = 2.37 cmolc kg^{-1} ; Na = 0.04 cmolc kg^{-1} ; K = 0.02 cmolc kg^{-1} ; H = 0.95 cmolc kg^{-1} ; Al = 0.20 cmolc kg^{-1} ; organic matter = 6.5 g kg^{-1} and P = 21.7 mg kg^{-1} . The pots were filled with previously air-dried, sieved soil and fertilized with 250 kg ha^{-1} of P2O5 in the foundation and 300 kg ha^{-1} of N in the top dressing, divided into six applications (sowing; 10; 20; 30; 40 and 50 days after sowing - DAS) using triple superphosphate and urea fertilizers, respectively. The treatments consisted of two doses of Cu (16 and 32 mg dm^{-3}) combined with two doses of Zn (32 and 64 mg dm^{-3}) and two doses of K_2O (150 and 250 kg ha^{-1}), arranged in a completely randomized design, in a 2 x 2 x 2 factorial scheme, with three replications, making a total of 24 experimental units for each experiment. The K_2O fertilizer, the source of which was potassium chloride, was spread over six applications, concomitantly with the N; the Cu and Zn fertilizers, the sources of which were copper sulphate and zinc sulphate, were spread over four applications (sowing; 20; 40 and 60 DAS). One experiment used the castor bean BRS Nordestina and the other BRS Paraguaçu; each pot received five castor bean seeds, keeping only one plant per pot, followed by thinning at 20 days after emergence. During the experimental period (123 days), the soil was kept at 90% of its field capacity, using a TDR probe to monitor the humidity on a daily basis. At 21, 42, 62, 82, 102 and 123 DAS, plant growth characteristics were assessed, such as: plant height, stem diameter at the base, number and length of leaves; the leaf area (LA) was calculated according to Wendt's method (1967), using the formula $\log(Y) = \Sigma\{-0.346 + [2.152 \times \log(X)]\}$, where Y is the leaf area in cm^2 and X is the length of the central vein of the leaf, in cm. At 123 DAS, the fruit was collected and submitted to an oven at 60° C with air circulation, until a constant weight was obtained; the seeds were then removed from the fruit, counted and weighed separately for each plant; finally, the data was submitted to analysis of variance (ANOVA) (Ferreira, 2011).

With the exception of the BRS Paraguaçu cultivar with the treatments that received the combination of Cu and Zn with 16 and 32 mg dm^{-3} and with 32 and 32 mg dm^{-3} (Figure 1B), the treatments with 150 kg ha^{-1} of K_2O caused greater heights than the plants submitted to the treatments with 250 kg ha^{-1} . It is likely that the higher dose of K may have caused greater osmotic stress, damaging the cultivar's development and production (Pacheco et al., 2008).

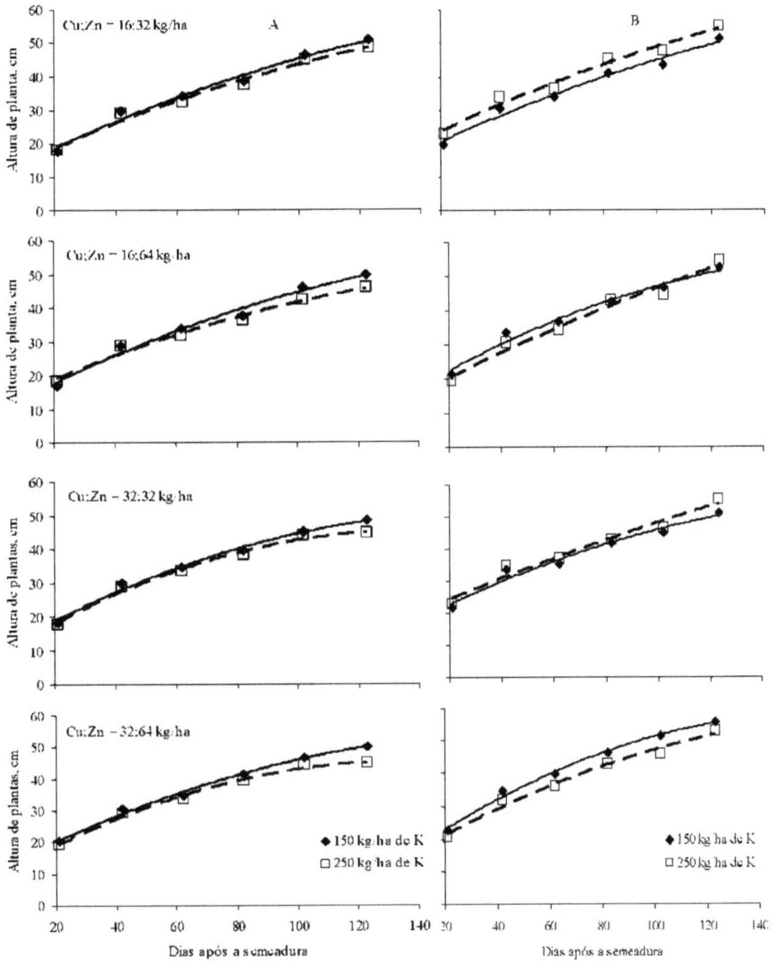

Figure 1. Plant height, cultivars BRS 149 Nordestina (A) and BRS 188 Paraguaçu (B), subjected to treatments with 150 and 250 kg ha^{-1} of K and dosages of Cu and Zn, respectively, in the following proportions: 16 and 32 kg ha^{-1} ; 16 and 64 kg ha^{-1} ; 32 and 32 kg ha^{-1} ; 32 and 64 kg ha^{-1} . Source: Dantas Junior et al. (2010)

The mathematical models of castor bean plant growth are shown in Table 1.

Table 1. Curve equations for the height of castor bean plants, cultivars BRS Nordestina and BRS Paraguaçu, subjected to treatments with 150 and 250 kg ha^{-1} of K and dosages of Cu and Zn, respectively, in the following proportions: 16 and 32 mg dm^{-3} ; 16 and 64 mg dm^{-3} ; 32 and 32 mg dm^{-3} ; 32 and 64 mg dm^{-3}

Treatment.	Plant height curve equations	
K Cu Zn	BRS Nordestina	BRS Paraguaçu
150-16-32	$y = -0.001 x^2 + 0.456 x + 9.903$ $R^2 = 0.98$	$y = -0.0007 x^2 + 0.392 x + 13.043$ $R^2 = 0.98$
250-16-32	$y = -0.001 x^2 + 0.429 x + 10.371$ $R^2 = 0.98$	$y = -0.0008 x^2 + 0.407 x + 15.843$ $R^2 = 0.98$
150-16-64	$y = -0.001 x^2 + 0.473 x + 8.668$ $R^2 = 0.98$	$y = -0.0012 x^2 + 0.458 x + 13.805$ $R^2 = 0.98$
250-16-64	$y = -0.001 x^2 + 0.408 x + 11.089$ $R^2 = 0.98$	$y = -0.0005 x^2 + 0.393 x + 12.71$ $R^2 = 0.97$
150-32-32	$y = -0.0016 x^2 + 0.517 x + 9.039$ $R^2 = 0.99$	$y = -0.001 x^2 + 0.415 x + 14.913$ $R^2 = 0.97$
250-32-32	$y = -0.002 x^2 + 0.557 x + 7.463$ $R^2 = 0.99$	$y = -0.0004 x^2 + 0.348 x + 17.557$ $R^2 = 0.97$
150-32-64	$y = -0.002 x^2 + 0.481 x + 11.342$ $R^2 = 0.99$	$y = -0.0017 x^2 + 0.561 x + 12.355$ $R^2 = 0.99$
250-32-64	$y = -0.002 x^2 + 0.546 x + 8.778$ $R^2 = 0.99$	$y = -0.0012 x^2 + 0.466 x + 12.326$ $R^2 = 0.98$

Source: Dantas Junior et al. (2010)

The plant height values of the two cultivars analyzed as a function of the potassium treatments at the different times were only significant at 123 DAS. Zinc treatments did not influence plant height; copper had a significant effect only at 62 DAS in the BRS Paraguaçu cultivar; the combination of these elements did not influence plant height in the BRS Nordestina cultivar, however, from 42 DAS onwards, the combination of potassium and zinc had a significant effect in the BRS Paraguaçu cultivar (Table 2).

Table 2. Summary of the analysis of variance for plant height at 21, 42, 62, 82, 102 and 123 days after sowing of the BRS Nordestina and BRS Paraguaçu cultivars, subjected to increasing doses of potassium, copper and zinc.

Treatments	Days after sowing					
	21	42	62	82	102	123
	Height of BRS Nordestina plants (cm)					
K	ns	ns	ns	ns	ns	*
Cu	ns	ns	ns	ns	ns	ns
Zn	ns	ns	ns	ns	ns	ns
K x Cu	ns	ns	ns	ns	ns	ns
K x Zn	ns	ns	ns	ns	ns	ns
Cu x Zn	ns	ns	ns	ns	ns	ns
K x Cu x Zn	ns	ns	ns	ns	ns	ns
M.G.	18,29	29,29	33,66	38,62	44,99	47,85
CV(%)	11,13	7,58	6,44	6,91	7,54	8,10
	Height of BRS Paraguaçu plants (cm)					
K	ns	ns	ns	ns	ns	*
Cu	ns	ns	*	ns	ns	ns
Zn	ns	ns	ns	ns	ns	ns
K x Cu	ns	ns	ns	ns	ns	ns
K x Zn	ns	*	**	*	**	*
Cu x Zn	ns	ns	ns	ns	ns	ns
K x Cu x Zn	ns	ns	ns	ns	ns	ns
M.G.	21,44	32,94	36,37	43,21	46,44	53,57
CV (%)	14,26	8,39	4,10	5,52	5,92	3,96

M.G = overall mean; CV = coefficient of variation, ns = not significant; * and ** = respectively significant at 5% and 1%. Source: Dantas Junior et al. (2010).

Throughout the experimental period, plant height values for the BRS Nordestina and BRS Paraguaçu cultivars ranged from 20 to 50 cm and 20 to 56 cm, with overall averages of 47.85 and 53.57 cm, respectively. These values were lower than those observed by Rodrigues et al. (2009), Araujo et al. (2009), Ribeiro et al. (2009) and (Chaves et al., 2009a).

The stem diameter is an important parameter in non-destructive growth analyses and, according to the results obtained, this parameter showed an increase over time (Figure 2), whose mathematical models are shown in Table 3.

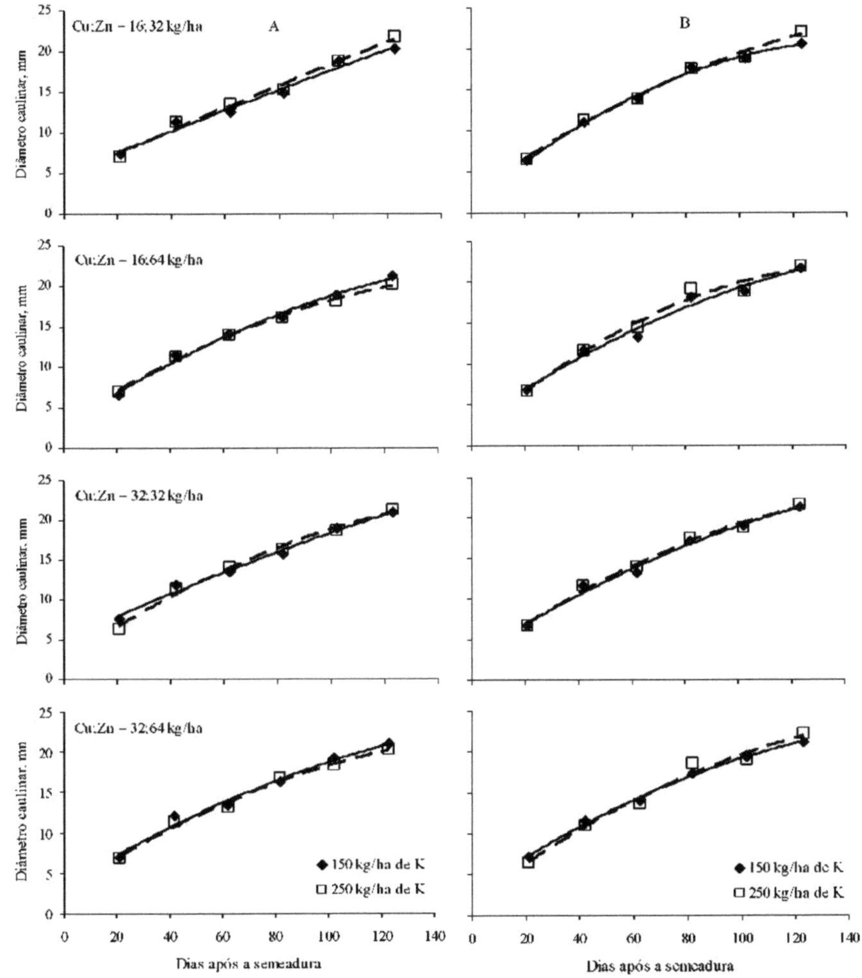

Figure 2. Culm diameter, cultivars BRS 149 Nordestina (A) and BRS 188 Paraguaçu (B), subjected to treatments with 150 and 250 kg ha^{-1} of K and dosages of Cu and Zn, respectively, in the following proportions: 16 and 32 kg ha^{-1} ; 16 and 64 kg ha^{-1} ; 32 and 32 kg ha^{-1} ; 32 and 64 kg ha^{-1}. Source: Dantas Junior et al. (2010)

Table 3. Equations of the stem diameter curves of castor bean plants, cultivars BRS Nordestina and BRS Paraguaçu submitted to treatments with 150 and 250 kg ha^{-1} of K and dosages of Cu and Zn, respectively, in the following proportions: 16 and 32 mg dm^{-3} ; 16 and 64 mg dm^{-3} ; 32 and 32 mg dm^{-3} ; 32 and 64 mg dm^{-3}

Treatment.	Equations for stem diameter curves	
K Cu Zn	BRS Nordestina	BRS Paraguaçu
150-16-32	$y = -7E{-}05\,x^2 + 0.136\,x + 4.835$ $R^2 = 0.98$	$y = -0.0009\,x^2 + 0.272\,x + 1.037$ $R^2 = 0.99$
250-16-32	$y = -6E{-}05\,x^2 + 0.145\,x + 4.565$ $R^2 = 0.99$	$y = -0.0006\,x^2 + 0.227\,x + 2.245$ $R^2 = 0.99$
150-16-64	$y = -0.0006\,x^2 + 0.218\,x + 2.606$ $R^2 = 0.99$	$y = -0.0008\,x^2 + 0.248\,x + 2.054$ $R^2 = 0.98$
250-16-64	$y = -0.0005\,x^2 + 0.207\,x + 3.159$ $R^2 = 0.99$	$y = -0.001\,x^2 + 0.283\,x + 1.142$ $R^2 = 0.98$
150-32-32	$y = -0.0002\,x^2 + 0.152\,x + 4.948$ $R^2 = 0.99$	$y = -0.0005\,x^2 + 0.208\,x + 2.872$ $R^2 = 0.99$
250-32-32	$y = -0.0006\,x^2 + 0.222\,x + 2.395$ $R^2 = 0.99$	$y = -0.0006\,x^2 + 0.231\,x + 3.389$ $R^2 = 0.99$
150-32-64	$y = -0.0005\,x^2 + 0.199\,x + 3.454$ $R^2 = 0.98$	$y = -0.0006\,x^2 + 0.223\,x + 2.868$ $R^2 = 0.99$
250-32-64	$y = -0.0006\,x^2 + 0.213\,x + 2.852$ $R^2 = 0.99$	$y = -0.0006\,x^2 + 0.242\,x + 1.743$ $R^2 = 0.98$

Source: Dantas Junior et al. (2010)

The treatments with K, Cu and Zn fertilizers, at the dosages and under the conditions in which the experiments were conducted, resulted in significant differences in stem diameters in some situations (Table 4).

Table 4 Summary of the analysis of variance for stem diameter at 21, 42, 62, 82, 102 and 123 days after sowing (DAS) of the BRS Nordestina and BRS Paraguaçu cultivars, submitted to potassium, copper and zinc treatments.

Treatments	Days after sowing					
	21	42	62	82	102	123
	Stem diameter of BRS Nordestina (mm)					
K	ns	*	ns	ns	ns	ns
Cu	ns	*	ns	ns	ns	ns
Zn	ns	ns	ns	ns	ns	ns
K x Cu	ns	ns	ns	ns	ns	ns
K x Zn	ns	ns	ns	ns	ns	ns
Cu x Zn	ns	ns	*	ns	ns	ns
K x Cu x Zn	ns	ns	ns	ns	ns	ns
M.G.	6,97	11,55	13,52	15,92	18,75	20,93
CV(%)	7,96	3,35	5,49	9,83	6,27	4,78
	Stem diameter of BRS Paraguaçu (mm)					
K	ns	ns	ns	ns	ns	*
Cu	ns	ns	ns	ns	ns	ns
Zn	ns	ns	ns	*	ns	ns
K x Cu	ns	ns	ns	ns	ns	ns
K x Zn	ns	ns	ns	ns	ns	ns
Cu x Zn	ns	ns	ns	ns	ns	ns
K x Cu x Zn	ns	ns	*	ns	ns	ns
M.G.	6,67	11,43	13,76	17,86	18,95	21,45
CV (%)	5,42	4,38	4,02	5,47	6,38	5,42

M.G = overall mean; CV = coefficient of variation, ns = not significant; * and ** = respectively significant at 5% and 1%. Source: Dantas Junior et al. (2010)

The average values obtained at 123 DAS were 20.93 mm and 21.45 mm for the BRS Nordestina and BRS Paraguaçu cultivars, respectively, lower than those observed by several authors (Araujo et al., 2009; Mesquita, 2010; Rodrigues et al., 2009; Silva et al. , 2008).

Leaf area is one of the most important indicators of plant growth because it reflects the size of the assimilatory apparatus, which is directly related to the plant's physiological processes.

In all treatments, leaf area increased until approximately 100 DAS; after this period, there was a decrease accompanied by leaf abscission, probably due to leaf senescence (Figure 3). The paralysis of vegetative growth due to the acceleration of reproductive growth occurs because photoassimilates are directed towards the growth of productive organs, considering that the fruit in formation acts as a drain on photoassimilates, thus causing a reduction in leaf area (Taiz and Zeiger, 2004).

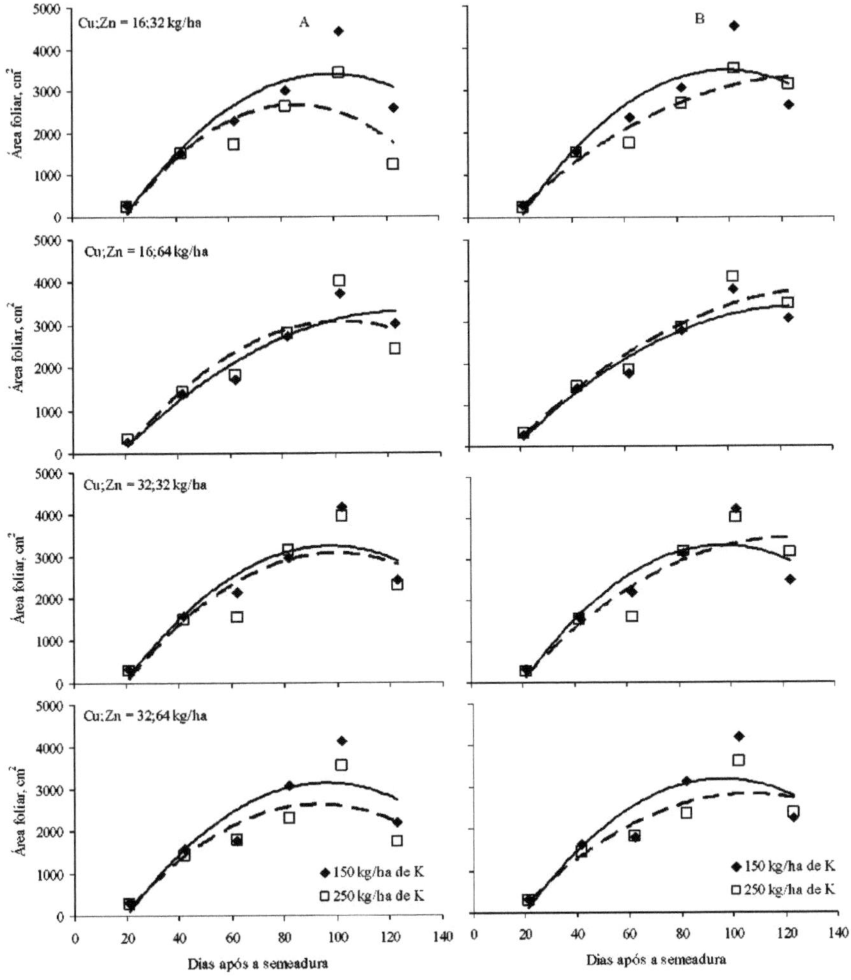

Figura 3. Leaf area, cultivars BRS 149 Nordestina (A) and BRS 188 Paraguaçu (B), submitted to treatments with 150 and 250 kg ha^{-1} of K and dosages of Cu and Zn, respectively, in the following proportions: 16 and 32 kg ha^{-1} ; 16 and 64 kg ha^{-1} ; 32 and 32 kg ha^{-1} ; 32 and 64 kg ha^{-1}. Source: Dantas Junior et al. (2010)

The mathematical models of the leaf area of the castor bean plants, cultivars BRS Nordestina and BRS Paraguaçu submitted to the treatments are shown in Table 5.

Table 5 - Equations of the leaf area curves of castor bean plants, cultivars BRS Nordestina and BRS Paraguaçu, submitted to treatments with 150 and 250 kg ha-1 of K and dosages of Cu and Zn, respectively, in the following proportions: 16 and 32 mg dm^{-3} ; 16 and 64 mg dm^{-3} ; 32 and 32 mg dm^{-3} ; 32 and 64 mg dm^{-3}

Treatment. K Cu Zn	Leaf area curve equations BRS Nordestina	BRS Paraguaçu
150-16-32	$y = -0.552x^2 + 108.95x - 1966.9$ $R^2 = 0.84$	$y = -0.5618x^2 + 110.85x - 2001.3$ $R^2 = 0.84$
250-16-32	$y = -0.645x^2 + 109.25x - 1955.4$ $R^2 = 0.75$	$y = -0.2541x^2 + 66.54x - 1031.1$ $R^2 = 0.95$
150-16-64	$y = -0.278x^2 + 70.826x - 1182.5$ $R^2 = 0.92$	$y = -0.2833x^2 + 72.07x - 1203.2$ $R^2 = 0.92$
250-16-64	$y = -0.447x^2 + 91.27x - 1574$ $R^2 = 0.82$	$y = -0.2452x^2 + 69.56x - 1100.9$ $R^2 = 0.93$
150-32-32	$y = -0.533x^2 + 103.67x - 1795.6$ $R^2 = 0.85$	$y = -0.5424x^2 + 105.49x - 1827.1$ $R^2 = 0.85$
250-32-32	$y = -0.490x^2 + 97.32x - 1753,6$ $R^2 = 0.79$	$y = -0.3261x^2 + 79.83x - 1371.9$ $R^2 = 0.88$
150-32-64	$y = -0.542x^2 + 103.79x - 1831.1$ $R^2 = 0.78$	$y = -0.5516x^2 + 105.61x - 1863.2$ $R^2 = 0.77$
250-32-64	$y = -0.482x^2 + 89.22x - 1496$ $R^2 = 0.76$	$y = -0.3595x^2 + 76.25x - 1210$ $R^2 = 0.85$

Source: Dantas Junior et al. (2010)

The treatments with K, Zn and the interaction between K and Zn had a significant effect on leaf area at 62 DAS for the BRS Nordestina cultivar; for the BRS Paraguaçu cultivar, at 62 DAS, leaf area was significantly influenced by K, Zn and the interaction between K and Zn and, at 123 DAS, by K, Cu and the interaction between Cu and Zn. However, the highest overall leaf area averages were at 102 DAS for the BRS Nordestina and BRS Paraguaçu cultivars, corresponding to 3921.22 and 3989.84 cm^2 , respectively (Table 6).

Table 6. Summary of the analysis of variance for leaf area at 21, 42, 62, 82, 102 and 123 days after sowing (DAS) of the BRS Nordestina and BRS Paraguaçu cultivars, submitted to potassium, copper and zinc treatments.

Treatments	Days after sowing 21	42	62	82	102	123
			Leaf area of BRS Nordestina (cm)²			
K	ns	ns	**	ns	ns	ns
Cu	ns	ns	ns	ns	ns	ns
Zn	ns	ns	*	ns	ns	ns
K x Cu	ns	ns	ns	ns	ns	ns
K x Zn	ns	ns	**	ns	ns	ns
Cu x Zn	ns	ns	ns	ns	ns	ns
K x Cu x Zn	ns	ns	ns	ns	ns	ns
M.G.	288,80	1492,62	1853,31	2835,92	3921,22	2243,74
CV(%)	20,63	8,69	8,75	20,94	19,25	42,90
			Leaf area of BRS Paraguaçu (cm)²			
K	ns	ns	**	ns	ns	*
Cu	ns	ns	ns	ns	ns	**
Zn	ns	ns	*	ns	ns	ns
K x Cu	ns	ns	ns	ns	ns	ns
K x Zn	ns	ns	**	ns	ns	ns
Cu x Zn	ns	ns	ns	ns	ns	**
K x Cu x Zn	ns	ns	ns	ns	ns	ns
M.G.	293,85	1518,72	1885,73	2885,57	3989,84	2816,29
CV (%)	20,63	8,69	8,75	20,94	19,25	12,71

M.G = overall mean; CV = coefficient of variation, ns = not significant; * and ** = respectively significant at 5% and 1%. Source: Dantas Junior et al. (2010)

In the BRS Nordestina cultivar, the K, Cu and Zn fertilization treatments did not result in statistically significant differences in seed number and weight parameters. The variation in nutrients under the conditions of the experiment did not lead to differences in the average yields (Table 7). It was observed that the number (ranging from 34 to 63) and weight (ranging from 10.53 to 20.42 g) of seeds in this experiment with 150 and 250 kg of K_2O ha^{-1} (Table 8) were lower than the 69 and 32.42 g of seeds, respectively, obtained in the experiment with the same cultivar fertilized with only 30 kg ha^{-1} of K_2O (Araujo, 2010). The number of seeds of the BRS Paraguaçu cultivar showed a significant difference at a 5% probability level when fertilized with potassium. The treatments fertilized with 250 kg ha^{-1} of K_2O had an average of 87.71 seeds, while the treatments fertilized with 150 kg ha^{-1} of K_2O had an average of 76.38 seeds. Another factor that caused a significant difference at the 1% probability level was copper fertilization, with averages of 90.79 and 73.30 seeds for the treatments fertilized with 16 and 32 mg dm^{-3} , respectively. The interaction between potassium and zinc was significant at the 5% probability level and the interaction between the other nutrients under study (K, Zn and Cu) did not significantly affect the number of seeds of the BRS Paraguaçu cultivar (Table 7). The seed weight of the BRS Paraguaçu cultivar showed a significant difference at the 1% probability level only for the interaction between the nutrients under study (K, Zn and Cu), and there were no statistical differences in castor bean seed production when the nutrients were evaluated separately (Table 7), corroborating Ribeiro et al. (2009).

Table 7. Mean squares of the analysis of variance for the number and weight of seeds of the BRS Nordestina and BRS Paraguaçu cultivars, subjected to treatments with K, Cu and Zn.

Variation factor	Mean squares		
	GL	No. of seeds	Seed weight, g
		BRS Nordestina	
R	2	635.79 ns	180.87 ns
K	1	28.17 ns	32.67 ns
Cu	1	140.17 ns	28.17 ns
Zn	1	80.67 ns	2.67 ns
K x Cu	1	560.67 ns	60.17 ns
K x Zn	1	748.17 ns	42.67 ns
Cu x Zn	1	4.17 ns	88.17 ns
K x Cu x Zn	1	112.67 ns	20.17 ns
Error	14	195,93	65,83
CV(%)		30,84	49,17
		BRS Paraguaçu	
R	2	422.10 ns	22.18 ns
K	1	769,53 *	133.95 ns
Cu	1	1835,75 **	140.65 ns
Zn	1	474.37 ns	53.10 ns
K x Cu	1	32.90 ns	31.05 ns
K x Zn	1	785,47 *	46.76 ns
Cu x Zn	1	0.004 ns	6.30 ns
K x Cu x Zn	1	469.05 ns	609,03 **
Error	14	152,76	52,58
CV (%)		15,06	17,55

CV = Coefficient of variation, ns = non-significant; * and ** = respectively significant at 5% and 1%. Source: Dantas Junior et al. (2010)

The combination of 250 kg ha^{-1} of K_2O with 16 and 32 mg dm^{-3} of Cu and Zn, respectively, was the one that produced the best results, resulting in average seed weights of 51.67 g; the other combinations had lower average seed weights (Table 8); however, these values were higher than the 34.21 g found by Ribeiro et al. (2009), using 150 kg ha^{-1} of K_2O, with the exception of the 31.81 g found for the treatment of 150 kg ha^{-1} of K_2O with 16 and 32 mg dm^{-3} Cu and Zn, respectively.

Table 8. Average number and weight of seeds of the BRS Nordestina and BRS Paraguaçu cultivars, fertilized with different doses of K, Cu and Zn.

Treatments	BRS Nordestina		BRS Paraguaçu	
K Cu Zn	No. of seeds	Seed weight, g	No. of seeds	Weight of Seeds, g
150-16-32	45	16,83	72	31,81
250-16-32	49	17,19	101	51,67
150-16-64	63	14,59	101	48,67
250-16-64	36	12,98	90	42,80
150-32-32	34	10,53	61	40,34
250-32-32	48	20,42	77	35,50
150-32-64	45	19,46	72	35,00
250-32-64	46	20,17	83	44,75

Source: Dantas Junior et al. (2010)

When working with different doses of fertilizer, a number of hypotheses, or even a combination of different factors, can explain the non-significance of treatments on plant growth parameters. According to the law of minima (RAIJ, 1991), vegetative growth is limited by the least available nutrient in the soil or even by increasing amounts of fertilizer, to the point of reaching levels that are detrimental to plant growth.

According to the above, the doses of potassium, copper and zinc influenced the growth of the BRS Paraguaçu and BRS Nordestina cultivars and the production of BRS Paraguaçu, with the highest seed weight production of BRS Paraguaçu corresponding to 250 kg ha^{-1} of K_2O, 16 and 32 mg dm^{-3} of Cu and Zn, respectively.

References

Araujo, D. L. N P K fertilization in the agronomic behavior of castor bean cultivar BRS 149 Nordestina in a protected environment. 2010. 75 p. Dissertation (Master's Degree in Agricultural Engineering)- Federal University of Campina Grande, Campina Grande, 2010.

Araujo, D. L.; Chaves, L. H. G.; Mesquita, E. F.; França, C. P. Growth of papaya tree cultivar BRS-149 Nordestina fertilized with nitrogen, phosphorus and potassium. Engenharia Ambiental, v.6, n. 3, p. 685-702, 2009.

Chaves, L. H. G.; Barros Junior, G.; Lacerda, R. D.; Cabral, P. C.; Cunha, T. H. C. S. Response of papaya fertilized with zinc and copper. Engenharia Ambiental, v. 6, n. 3, p. 306-319, 2009a.

Chaves, L. H. G.; Barros Junior, G.; Lacerda, R. D.; Cabral, P. C.; Cunha, T. H. C. S. Effect of zinc and copper on the nutritional status of the papaya tree, BRS 188 Paraguaçu. Caatinga, v. 22, n. 4, p.

129-135, 2009b.

Dantas Junior, E. E.; Chaves, L. H. G.; Costa, F. A. M.; Mesquita, E. F.; Araujo, D. L. Growth of two papaya cultivars fertilized with potassium, copper and zinc. Revista Caatinga, v.23, n. 4, p. 97-107, 2010.

Ferreira, D.F. Sisvar: a computer statistical analysis system. Ci. Agrot., v. 35, p.1039-1042, 2011.

Mesquita, E. F. Agronomic behavior of two castor bean cultivars under mineral fertilization. 2010. 111p. Thesis (Doctorate in Agricultural Engineering) - Federal University of Campina Grande, Campina Grande, 2010.

Pacheco, D. D.; Gonçalves, N. P.; Saturnino, H. M.; Antunes, P. D. Production and availability of nutrients for papaya (Ricimus communis L.) fertilized with NPK. Revista de Biologia e Ciências da Terra, v. 8, n. 1, p. 153-160, 2008.

Raij, B. van. Soil fertility and fertilization. Piracicaba: Agronòmica Ceres, 1991. 343p.

Ribeiro, S.; Chaves, L. H. G.; Guerra, H. O. C.; Gheyi, H. R.; Lacerda, R. D. Response of the papaya cultivar BRS-188 Paraguaçu to the application of nitrogen, phosphorus and potassium. Revista Ciência Agronòmica, v. 40, n. 4, p. 465-43, 2009.

Rodrigues, L. N.; Nery, A. R.; Fernandes, P. D.; Beltrao, N. E. M.; Gheyi, H. R. Growth and berry production of papaya irrigated with domestic wastewater. Revista Brasileira de Engenharia Agricola e Ambiental, v. 13 (Suplemento), p. 825-853, 2009.

Silva, S. M. S.; Alves, A. N.; Gheyi, H. R.; Beltrao, N. E. M.; Severino, L. S.; Soares, F. A. L. Development and production of two cultivars of papaya under salt stress. Revista Brasileira de Engenharia Agricola e Ambiental, v. 12, n. 4, p. 335-342, 2008.

Taiz, L.; Zeiger, E. Plant physiology. 3.ed. Porto Alegre: artmed, 2004. 719p.

Wendt, C. W. Use of a relationship between leaf length and leaf area to estimate the leaf area of cotton *(Gossypium hirsutum* L.), castor *(Ricimus communis* L.) and sorghum (*Sorghum vulgare* L.). (1967). Agronomy Journal, v.59, n.5, p.485-487, 1967.

CHAPTER 3

NITROGEN FERTILIZATION VERSUS WATER CONSUMPTION

All plants need water and nutrients for their development, some more, some less depending on the species of plant. In the case of the castor bean plant, it has been observed that it produces economically viable crops with an amount of water that is considered insufficient for other crops. The ideal rainfall range for castor bean production varies between 750 and 1500 mm, with a minimum of 600 to 750 mm throughout the crop cycle, with planting adjusted so that the plant receives 400 to 500 mm by the start of flowering (Tàvora, 1982). In regions with rainfall totals of less than 500 mm during the rainy season, the castor bean plant loses a large part of its economic production, increasing the risk of total crop loss and/or very low yields. An excess of water in the soil impairs the development of castor beans, causing considerable damage to the physiology and productivity of the crop. Fertilization, in turn, is one of the main technologies used to increase the productivity and profitability of the crop, with nitrogen (N) being one of the main macronutrients responsible for this increase (Case study 6). For this reason, studies were carried out on nitrogen fertilization versus water consumption and water use efficiency in castor bean cultivation, which are presented below.

CASE STUDY 5

The objective of Case Study 6 was to evaluate the effect of nitrogen fertilization on phytomass production, water consumption and water use efficiency of the BRS 188 Paraguaçu castor bean cultivar, and the cultivar's ability to reverse the volume of water consumed in dry matter production. To this end, an experiment was carried out in a greenhouse belonging to the Department of Agricultural Engineering at the Federal University of Campina Grande, using plastic pots with a capacity of 100L, perforated at the base to allow drainage, which were filled with 78 kg of sandy loam soil material (typical Eutrophic Regolithic Neosol). The treatments consisted of five levels of nitrogen (40; 80; 120; 160 and 200 kg ha^{-1}) with three replications. In addition, the plants received 90 and 60 kg ha^{-1} of P2O5 and K2O, using urea, triple superphosphate and potassium chloride fertilizers, respectively.

During the experimental period (140 days), one plant was kept in each pot with humidity at 80% of field capacity, monitored daily using a TDR probe - Delta T model HH2. Irrigations were carried out every time the water content in the soil reached less than 80% of field capacity, and the water was replaced immediately after each reading. At 140 days after planting (DAP), the weight of the dry phytomass, the water consumption of each plot in volume and the cumulative consumption were evaluated, so that the efficiency of this consumption was determined by the ratio between the weight of the total dry matter (stem, branches, leaves, fruit and roots) and the volume of water actually consumed in each treatment, according to the methodology described by Barker et al. (1989). In order to analyze the effect of the nitrogen treatments, the experimental data was analyzed using analysis of variance (ANOVA), Tukey's test at 5% probability and regression analysis, according to Ferreira (2011).

The analysis of variance showed that the treatments had a significant effect at 5% probability on total phytomass and use efficiency. However, not only these two parameters, but also water consumption by the plants, increased as a result of the higher doses of nitrogen applied to the soil (Table 1).

Table 1. Averages of phytomass data, total water consumption and water use efficiency and linear equations of the parameters.

Doses of N $(kg\ ha)^{-1}$	Total phytomass (g)	Water consumption (L)	Efficiency $(g\ L)^{-1}$
40	133,74	91,28	1,46
80	158,70	98,71	1,61
120	215,70	112,56	1,92
160	209,17	114,11	1,83
200	285,44	124,55	2,29
Total phytomass	$Y = 94.389 + 0.8847\ x$		$R^2 = 0.91$
Water consumed	$Y = 83.66 + 0.2049\ x$		$R^2 = 0.96$
Efficiency	$Y= 1.258 + 0.0047\ x$		$R^2 = 0.87$

The efficiency of water use increased with the doses of N applied to the soil, and the highest efficiency, 2.29 g L^{-1} , was observed with the highest nitrogen treatment, i.e. the plants were better able to convert the volume of water consumed into dry matter production whenever the nitrogen level was increased. According to the efficiency equation, a gain of 0.042 g L^{-1} was obtained for each unit increase in the percentage of water available in the soil.

According to the results, the castor bean cultivar BRS 188 Paraguaçu increased its phytomass, water consumption and water use efficiency with increasing doses of nitrogen applied to the soil, indicating high efficiency in transforming the water consumed into dry matter.

References

Barker, R. E.; Frank, A. B.; Berdahl, J. D. Cultivar and clonal differences for water use efficiency and yield in four forage grasses. Crop Science, v. 29, p. 58-61, 1989.

Chaves, Gheyi, H. R.; Ribeiro, S. Water consumption and use efficiency for castor bean cultivar Paraguaçu submitted to nitrogen fertilization. Engenharia Ambiental, v. 8, n. 1, p. 126-133, 2011.

Ferreira, D.F. Sisvar: a computer statistical analysis system. Ci. Agrot., v. 35, p.1039-1042, 2011.

CHAPTER 4

ORGANIC FERTILIZATION FOR CASTOR BEANS

Soils in the semi-arid region generally have low levels of organic matter, which is why it is essential to incorporate organic compounds into the soil. These compounds can cause changes in the physical, chemical and biological characteristics of soils, improving their structure, increasing water retention capacity, aeration and soil fertility (Malavolta et al., 1997). This can also promote greater growth and development of crops, such as papaya, as observed by Oliveira et al. (2009). However, it is worth pointing out that the successful use of organic materials in a crop will depend, among other factors, on the quality and quantity of the material to be used.

The high cost of mineral fertilizers for castor bean planting, the financial difficulties faced by farmers in the Northeast and the possibility of them producing organic compost with materials from their own properties, have led to several experiments being set up to evaluate the effect of different sources of organic fertilizer on the growth, development and production of castor bean.

CASE STUDY 6

In order to evaluate the vegetative growth of castor bean plants (Ricinus communis L.), cv. BRS Nordestina, subjected to organic and mineral fertilization, an experiment was set up in field and rainfed conditions on the Jacarè rural property located in the municipality of Remigio-PB (6°53'00"S; 36°02'00"O, altitude 470 m) with the following treatments: T1- absolute control (soil in its natural fertility condition); T2 - Garbage compost I (12 kg/well); T3 - Garbage compost II (12 kg/well of compost enriched with potassic and phosphatic rock dust, tile dust and ash); T4 - Mineral fertilization (in the foundation: 40 g of N; 177.77 g of P; 26.66 g of K; 22.857 g of Zn; 82.05 g of Mg; 22.598 g of B; 16 g of Cu; 14.28 g of Mn, per hole; 45 days after planting, 40 g of N/hole was fertilized); T5 - Organic fertilization with tanned farmyard manure. The cattle manure and waste composts were chemically characterized (Table 1).

Table 1. Chemical characteristics of the waste compost, cattle manure used in the experiment and the soil in the experimental area.

Fertilizers organic	Chemical characteristics							
	N	P	K	In	Ca	Mg	S	Cl
			g kg^{-1}----------					-
Manure	8,23	2,7	11,95	0,74	10,56	3,86	0,45	5,49
Waste compost I	9,12	4,82	7,05	0,89	17,61	2,67	0,45	5,6
Waste compost II	8,92	7,37	7,05	1,01	21,13	3,37	0,42	3,83
		Zn	Fe	Mn	Cu			
			mg kg-1----------------------					
Manure		44,93	3872	167	10,82			
Waste compost		73,23	8467	246	18,19			

I								
Waste compost II		75,05	10590	265	18,19			
	pH	Ca	Mg	In	K	Al	MO	P
Soil	(H2O)	cmol$_c$ kg^{-1} -------------------					g kg^{-1}	mg kg^{-1}
	6,17	1,12	0,85	1,38	4,29	0,16	11,0	7,7

Source: Fernandes et al. (2009).

Each experimental plot consisted of 4 rows with 4 plants/row, totaling 16 plants at a spacing of 2m between rows and 2m between plants, considering a useful area of 4 plants, in an area of 16m^2 . At 30, 90, 150 and 210 days after planting (DAP), the following parameters were assessed: height; stem diameter and number of leaves per plant. The seeds collected were dried in a greenhouse for 120 days. After this period, the oil was extracted from the seeds of each treatment by pressing. The oil was characterized in duplicate according to AOCS standards[3] (American Oil *Chemists Society), in* terms of acidity index, saponification index, iodine index, peroxide index, density, viscosity and moisture content. The results were submitted to analysis of variance and regression (Ferreira, 2011).

The soil in the experimental area, interpreted according to Lopes and Guidolin (1989), is weakly acidic, with a medium cation exchange capacity (CEC) (4.6 to 10 cmolc kg^{-1}), low levels of organic matter (0 to 15 g kg^{-1}) and available phosphorus (0 to 20 mg dm^{-3}). With the exception of the number of leaves at the end of the experiment (210 DAP), plant height (AP), stem diameter (DC) and number of leaves (NF) over the time evaluated were significantly influenced at a 1% probability level by the treatments, indicating that the castor bean plant is responsive to fertilizers, both organic and mineral. Treatments T2 and T3 (garbage composts I and II, respectively) did not differ statistically from each other over the same time intervals, and were the ones that generally showed the highest values for all the variables analyzed. The lowest values were obtained with T1 and T4 (control and mineral fertilizer, respectively) (Table 2).

Table 2. Breakdown of the fertilizer x crop interaction for the variables plant height, stem diameter and number of leaves for the period from 0 to 210 days after planting (DAP). ___

Parameters analyzed	Treatment averages T1	T2	T3	T4	T5
			30 DAP		
Plant height, cm	18,0c	37.2ab	41,7a	7,1d	31,5b
Stem diameter, cm	0,6c	1.3ab	1,5a	0,3c	1,1b
Sheet number	4,4b	6,0a	6,1a	2,0c	5.0ab
Plant height, cm	80,0c	120,3a	90 DAP 127.0a	74,0c	100,0b
Stem diameter, cm	1,3c	3,2a	3,6a	2,0b	3,5a
Sheet number	11,7b	24,8a	27,9a	13,4b	24,1a
			150 DAP		
Plant height, cm	144,7c	177.0ab	185.0a	146,0c	169,0b
Stem diameter, cm	2,8b	4,2a	5,1a	4,3a	4,9a
Sheet number	17,4c	44.2ab	55,7a	25.4bc	49,8a
			210 DAP		
Plant height, cm	152,0d	186.0ab	197,0a	168,5c	182.0bc
Stem diameter, cm	2,9b	4,2a	5,2a	4,3a	5,0a
Sheet number	14,3a	37,1a	41,7a	25,1a	43,1a

Means followed by the same letter in the rows are not statistically different. Source: Fernandes et al. (2009a)

It can be seen that the development of the plants that received no fertilizer (Control T1), up to 30 days after planting, was better than those that received mineral fertilizer (T4) (Table 2). The increase in the saline concentration in the planting hole during this period, due to the application of fertilizers, probably influenced the initial development of the plants. However, from then on, the lack and/or scarcity of nutrients in the control plant caused the plants to develop less. In general, the organic and/or mineral fertilization of the papaya tree led to an increase in its development, and the crop responded better to organic fertilizers, including waste compost. This was probably due not only to the increase in fertility, since the organic fertilizers added more nitrogen and potassium to the soil, but also to the improvement in the soil's physical properties, especially with regard to increased water retention.

The averages of the physico-chemical analyses of the crude castor oil as a function of the treatments (Table 3) are close to the standard established by the AOCS (Table 4), cited by Freire (2001).

Table 3. Physico-chemical analysis of castor oil as a function of fertilizer sources.

Treatments					DMS
Tl	T2	T3	T4	T5	
Acidity mg KOH/g					
0,69a	0,83a	0,69a	1,39b	0,69a	0,51
Yield %-------------					
31,97b	42,50a	44,37a	41,76a	42,47a	3,01
Density kg/m^3------					
0,962a	0,962a	0,962a	0,962a	0,962a	0,001
Viscosity mm^2 /s^2 100°C--					
19,99a	18,85a	19,55a	19,53a	19,19a	1,20
iodine index g I√100g----					
86,84a	91,69a	89,38a	89,79a	87,58a	18,80
peroxide index meq/1000g--					
0,0a	0,0a	0,0a	0,41a	0,0a	1,05
saponification index mg KOH/g---					
179.3ab	174,5a	175.5a187	.3b	181.0ab	1,94
Humidity %----------------					
0,56b	0.52ab	0.53ab	0,47a	0,58b	0,082

Averages followed by the same letter do not differ according to the Tukey test (P≤0.05). Source: Fernandes et al. (2009b).

Table 4: Castor oil specifications according to data from the Association of Official Analytical Chemists - AOAC.

Water content	acidity index	saponification index	iodine index	peroxide value
----%----	mg KOH/g---------		g I/100 g	meq/1000g
-	4 max.	176-187	81-91	-

Small variations, although not significant, between the means of the parameters viscosity, iodine index and peroxide index as a function of the treatments can be seen in Table 3. The non-significance of the iodine index shows a similarity between the oils, regardless of the fertilizer source used, in terms of the oil's unsaturation levels.

Viscosity values at 100°C ranged from 18.85 to 19.99 mm^2 /s^2 for treatments T2 and T1, respectively. High viscosity values are an advantage in the manufacture of lubricants, but, on the other hand, they are a problem to be overcome in the production of biodiesel, as the viscosity of this fuel has to fall within the limit specified by the ANP (2001). The highest acidity value was observed in treatment 4 (1.39), which differed statistically from the others. However, in all treatments the values were below the maximum permitted limit, showing that the seeds were of good quality. According to Santos et al. (2001) oil with an acidity of less than 1% is commercially classified as type 1 industrial oil and, when the oil has a maximum of 3% free acidity, it is recognized as type 3. The low acidity indices also show the oil's potential in the cosmetics industry. The lowest saponification index values were obtained using T2 and T3, both of which differed from T4 at 5% probability. However, all the values are close to those found in the literature. According to Chierice and Claro Neto (2001), castor oil has an average of 180 mg KOH/g. According to Costa et al. (2004), when studying castor oil they found a saponification index variation of 176 - 184 mg KOH/g. T4 had the lowest moisture content (0.47%) and was not statistically different from T2 and T3, which in turn were not different from T1. According to Santos et al. (2001) the castor oil obtained in T4 can be classified commercially as industrial type 1, as it is below the maximum permitted limit of 0.5% moisture.

References

ANP, Agência Nacional do Petróleo, Ministério de Minas e Energia, Portaria ANP n° 309 de 27/12/2001 Available at: https://www.legisweb.com.br/legislacao/?id=182588. Accessed on: March 12, 2009 Chierice, G.O.; Claro Neto, S. Aplicaçao industrial de óleo. In: Azevedo, D.M.P.;

Lima, E.F. (eds). The castor bean agribusiness in Brazil. Brasilia: EMBRAPA Technological Communication Service, 2001.350p.

Costa, H.M.; Ramos, V.D.; Abrantes, T.A.S.A.; Castro, L.L.Y.; Visconde, L.L.Y.; Nunes, R.C.R.; Furtado, C.R.G. Polimeros: Ciência e Tecnologia, v.14, n.1, p.46-50, 2004.

Fernandes, J. D.; Chaves, L. H. G.; Dantas, J. P.; Silva. J. R. P. Adubaçao orgânica e mineral no desenvolvimento da mamoneira. Engenharia Ambiental, v.6, n.2, p. 358-368, 2009a.

Fernandes, J. D.; Chaves, L. H. G.; Dantas, J. P.; Pinto, L. M.; Souza, A. G.; Cordeiro, A. Silva. J. R. P. Characterization of castor oil (Ricinus communis L.) grown using different sources of fertilizer. In: Congress of the Brazilian Biodiesel Technology Network, 3, 2009b. P. 439-440.

Ferreira, D.F. Sisvar: a computer statistical analysis system. Ci. Agrot., v. 35, p.1039-1042, 2011.

Freire, R. M. M. Ricinoquimica. In: Azevedo, D. M. P.; Lima, E. F. (Ed.). Castor bean agribusiness in Brazil. Brasilia: Embrapa Informaçao Tecnológica, 2001. p.295-335.

Lopes, A. S.; Guidolin, J. A. Interpreting soil analysis: concepts and applications. 2 ed. Sao Paulo: ANDA, 1989. 64p.

Santos, R.F.; Barros, M.A.L.; Marques, F.M.; Firmino, P.T. Requiao, L.E; G. Anàlise Econòmica In: Azevedo, D. M. P.; Lima, E. F. (eds). The castor bean agribusiness in Brazil. Brasilia: EMBRAPA technological communication service, 2001. 350p.

CASE STUDY 7

Some work has already been carried out to evaluate the influence of different sources of fertilizer on the development of castor beans (Ricinus communis L.), BRS 149 Nordestina and BRS 188 Paraguaçù (Fernandes et al., 2009; Santos et al., 2010), however, the need for studies and information on organic fertilization in more modern cultivars, such as cv . BRS Energia. This is why this case study shows the influence of different organic fertilizers on the growth of this cultivar. This case study also shows the importance of organic fertilizer in castor bean cultivation.

The experiment was set up in a greenhouse belonging to the Department of Agricultural Engineering at the Federal University of Campina Grande, using a randomized block design with four replications, in a 6 x 2 factorial scheme, with the first factor consisting of different fertilizer sources: F1 = absolute control; F2 = humus (12 kg/pot); F3 = cattle manure (12 kg/pot); F4 = humus + cattle manure (24 kg/pot of the 1:1); F5 = mineral fertilizer (40 g of N; 177.77 g of P2O5 and 26.66 g of K2O/pot) and F6 = supermagro biofertilizer (200 mL diluted in 1800 mL of water/pot), and the second factor, the absence and presence of coconut bagasse (400 g), totaling 12 treatments. Each plot consisted of 50 kg pots of Neossolo Regolitico containing just one plant, whose soil had the following characteristics: pH (H_2O) = 6.8; Ca = 5.05 cmolc kg^{-1} ; Mg = 3.02 cmolc kg^{-1} ; Na = 0.02 cmolc kg^{-1} ; K = 0.21 cmolc kg^{-1} ; H = 1.75 cmolc kg^{-1} ; Al = 0.00 cmolc kg^{-1} ; MO = 15.3 g kg^{-1} ; P = 37.8 mg kg^{-1} . The papaya tree was irrigated every 3 days with a 15.92 mm water line; as for the biofertilizer, it was applied at equidistant periods of 10 days, starting at 15 days after the seedlings emerged. At 96 days after planting, the following parameters were assessed: plant height up to the first raceme, stem diameter at the base, number of internodes up to the last leaf and leaf area according to the Wendt method (1967), using the formula Log (Y) = - 0.346 + [2.152 x Log (X)], where Y is the leaf area in cm^2 and X is the length of the central vein of the leaf in cm. The relationship between the length occupied by the female and male flowers in the raceme and the length of the raceme was also evaluated. The data was analyzed using SISVAR software (Ferreira, 2011) and the results were compared using Tukey's test.

Fertilizer sources significantly influenced plant height, stem diameter, raceme length and the ratio of female to male flowers (Table 1). Contrary to this, the use of coconut pomace had no significant effect on these parameters; however, there was an interaction between fertilizer source (F) and coconut pomace (B) for the number of internodes. The leaf area was not influenced by any factor; on average the castor bean plants had an area of 58,526.68 cm^2 in the absence and presence of coconut bagasse, respectively.

Table 1. Summary of the analysis of variance for plant height (AP), stem diameter (DC), leaf area (AF), number of internodes (NI), raceme length (CRAC) and ratio of female to male flowers (RF/M), subjected to different fertilizer sources and the presence and absence of coconut bagasse.

FV	Mean square					
	AP	DC	AF	NI	CRAC	RF/M
Fertilizer source (F)	2.796,4**	0,50**	2300685873,81^{ns}	4,43 ns	638,3**	2,64*
Bagasse (B)	102,9^{ns}	0,007^{ns}	6620314,21^{ns}	4,68^{ns}	93,2^{ns}	0,15^{ns}
F x B	239,6^{ns}	0,03^{ns}	1161085976,9^{ns}	5,43*	72,2^{ns}	1,17^{ns}
Treatment	1.389,4**	0,24**	1574134506,1^{ns}	4,91*	331,4**	1,74^{ns}
Block	596,5*	0,013^{ns}	1236300390,5^{ns}	0,68^{ns}	81,3^{ns}	0,56^{ns}
Waste	171,7	0,02	1567437187,8	1,89	32,0	1,05
CV (%)	12,5	8,40	66,92	10,5	26,9	60,67

$^{**, *, ns}$ Significant at 1 and 5% probability and not significant, respectively. Source: Chaves et al. (2011)

In the absence of coconut bagasse, the lowest number of internodes was found in the papaya trees fertilized with mineral fertilizer (F5), with the average differing only from the one that received worm humus (F2) (Table 2). With the addition of coconut bagasse, the number of internodes was statistically

similar regardless of the fertilizer source used. However, when the influence of coconut bagasse was analyzed among the sources, there was a reduction in NI when coconut bagasse was mixed with the fertilizers earthworm humus (F2) and cattle manure (F3).

Table 2. Breakdown of the interaction between fertilizer sources and coconut bagasse for the variable number of internodes (NI).

Variable	Bagasse	Fertilizer source					
		1	2	3	4	5	6
NI	Absence	13 abA	15 aA	14 abA	13 abA	12 bA	14 abA
	Presence	12 aA	13 aB	12 aB	14 aA	13 aA	14 aA

Averages followed by the same letter, lower case in the row and upper case in the column, do not differ statistically by the Tukey test. Source: Chaves et al. (2011)

With the exception of supermagro biofertilizer (F6), the organic sources, when compared to mineral fertilizer (F5), promoted the highest averages of plant height to first raceme (AP), stem diameter (DC), raceme length (CRAC) and ratio of female to male flowers (RF/M) (Table 3).

Table 3 Averages of plant height (AP), stem diameter (DC), raceme length (CRAC) and ratio of female to male flowers (RF/M) subjected to different fertilizer sources.

	Variable			
Fertilizer Source	AP(cm)	DC (cm)	CRAC (cm)	RF/M
Absolute witness	112,51 a	1.75 ab	26.62 ab	1.85 ab
Hùmus	111,92 a	1,88 a	20.47 bc	1.78 ab
Bovine manure	121,01 a	1,89 a	29,15 a	2,44 a
Hùmus + Bovine manure	118,9 a	1,96 a	28.91 ab	2.0 ab
Mineral fertilization	78,6 b	1,62 b	13.28 cd	0,81 b
Supermagro biofertilizer	82,63 b	1,28 c	7,63 d	1.26 ab

Source: Chaves et al. (2011)

As the number of internodes was not significantly influenced by the fertilizer source factor (Table 1), the higher plant height averages obtained with the use of organic fertilizers indicate that they have longer internodes; this information is important since in good environmental conditions (water and nutrients), internodes are long and in periods of drought or adverse conditions, internodes are short (Severino et al., 2007). The number of female flowers is the main objective of studies on sexual expression in papaya, as its reduction causes lower productivity in commercial plantations (Leles et al., 2010). According to Table 3, the highest RF/M ratios corresponded to the organic fertilizers.

References

Chaves, L. H. G.; Fernandes, J. D.; Cruz, M. P.; Brito, L.; Monteiro Filho, A. F.; Azevedo, M. R. Q. A. Effect of different fertilizer sources on the production of castor bean cultivar BRS Nordestina. Cadernos de Agroecologia, v. 6, n. 2, p. 1-5, 2011.

Fernandes, J. D.; Chaves, L. H. G.; Dantas, J. P.; Silva. J. R. P. Adubaçao orgânica e mineral no desenvolvimento da mamoneira. Engenharia Ambiental, v.6, n.2, p. 358-368, 2009.

Ferreira, D.F. Sisvar: a computer statistical analysis system. Ci. Agrot., v. 35, p.1039-1042, 2011.

Leles, E. P.; Fernandes, D. M.; Silva, J. I. C.; Fumes, L. A. A. Interaction between doses of lime and zinc on morphological characteristics and production components of papaya. Acta Scientiarum. Agronomy, v. 32, n. 3, p. 501-509, 2010.

Santos, J. F.; Grangeiro, J. I. T.; Oliveira, M. E. C. Production of papaya as a function of fertilization with chicken litter. Environmental Engineering: Research and Technology. Espirito Santo do Pinhal, v. 7, n. 1, p. 169180, 2010.

Severino, L. S.; Lucena, A. M. A.; Sampaio, L. R.; Tavares, M. J. V.; Beltrao, N. E. M.; Bortoluzi, C. R. D. Descriçao das Fases do Desenvolvimento Reprodutivo da Mamoneira Visando ao Manejo da Harvest. Campina Grande: Embrapa Cotton, 2007. 7p. (Technical circular, 115).

Wendt, C. W. Use of a relationship between leaf length and leaf area to estimate the leaf area of cotton *(Gossypium hirsutum* L.), castor *(Ricimus communis* L.) and sorghum (*Sorghum vulgare* L.). (1967). Agronomy Journal, v.59, n.5, p.485-487, 1967.

CASE STUDY 8

The aim of this case study was to evaluate the production components of cv BRS Nordestina when subjected to organic fertilization. To this end, an experiment was carried out under field and rainfed conditions on the Umbuzeiro rural property located in the municipality of Santa Luzia-PB, at an altitude of 304 m. Soil samples collected in the arable layer (0-20 cm) of the experimental area were chemically characterized according to the Embrapa methodology (1997), with the following results: pH ($_{H2O}$) = 6.37 ; Ca = 1.24 cmolc.kg^{-1} ; Mg = 0.49 cmolc.kg^{-1} ; Na = 0.04 mg.100g^{-1} ; K = 1.6 mg kg^{-}

^{1} ; Al = 0.28 cmolc.kg ;$^{-1}$

organic matter = 4.3 g kg^{-1} P = 8.5 mg kg^{-1} . The experimental design was randomized blocks with four replications and four treatments consisting of: F1- Absolute control (soil in its natural fertility condition); F2- Organic fertilization with tanned farmyard manure (12 kg/plot); F3- Mineral fertilization (in foundation: 40g of N; 177.77g of P_2O_5; 26.66g of K_2O; 22.857g of Zn; 82.05g of Mg; 22.598g of B; 16g of Cu; 14.28g of Mn, per hole; 45 days after planting, fertilization was carried out applying 40g of N/hole); F4- Organomineral fertilization (half of treatment F2) + mineral fertilization (half of treatment F3). The bovine manure was chemically characterized by Embrapa (1997), with the following results: N = 25.69 g kg^{-1} ; P = 4.89 g kg^{-1} ; K = 14.4 g kg^{-1} ; Na = 1.14 g kg^{-1} ; Ca = 17.61 g kg^{-1} ; Mg = 5.97 g kg^{-1} ; S = 0.91 g kg^{-1} ; Cl = 2.38 g kg^{-1} ; Zn = 71.88 mg kg^{-1} ; Fe = 2064 mg kg^{-1} ; Mn = 730 mg kg^{-1} ; Cu = 16.72 mg kg^{-1} . Seeds of the BRS Nordestina cultivar were planted in pits measuring 0.4 x 0.4 x 0.4 m. Each experimental plot consisted of four rows, with four plants/row, totaling 16 plants at a spacing of 2 m between rows and 2 m between plants. The four central plants of the plot were considered the useful area for the yield evaluations, covering an area of 16m^2 Taking into account the entire yield of all the raceme orders, the mass of the raceme per plant, the mass of grains per plant, the number of racemes per plant, the number of fruits per raceme and the percentage yield of grains were evaluated using the expression:

$$RPG = \frac{PT}{PTF} x100$$

Where: RPG = Percentage grain yield (%); PT = Total grain yield (kg plant^{-1}); PTF = Total fruit yield (kg plant);$^{-1}$

After collecting the data, it was submitted to analysis of variance using the "F" test and when a significant effect was found, the Tukey test was applied to compare means using the SISVAR statistical software (Ferreira, 2011).

On average, the highest raceme yields were seen with the application of cattle manure (F2) and organo-mineral fertilizer (F4), with 252.74 and 274.88 g respectively (Figure 1). The raceme mass obtained from plants fertilized with mineral fertilizer (F3) also did not differ statistically from those grown with cattle manure (F2) and in the absence of fertilizer (F1). According to Fernandes et al. (2010), the production of racemes with the use of cattle manure was, on average, 91.67% higher than with mineral fertilizer.

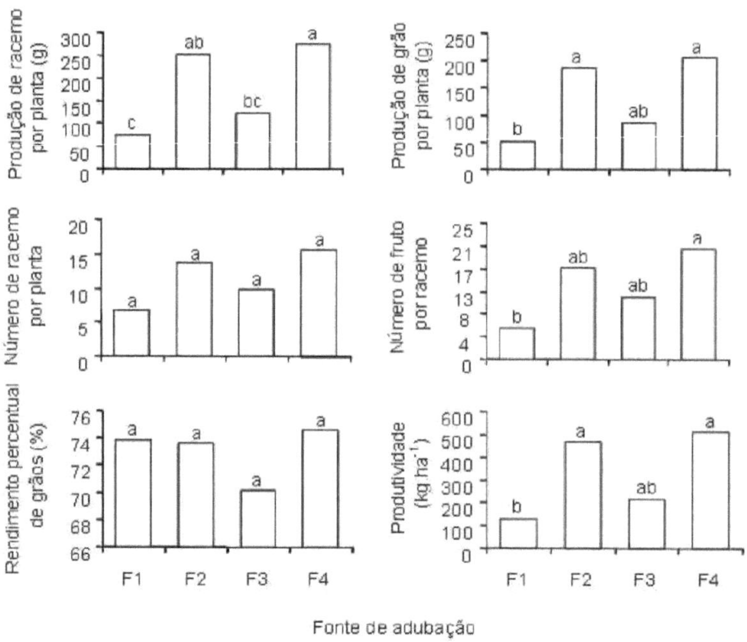

Figure 1. Yield components of BRS nordestina castor bean grown without fertilizer (F1), cattle manure (F2), mineral fertilizer (F3) and organomineral fertilizer (F4). source: chaves et al. (2011)

As for grain yield per plant, the use of bovine manure (F2) and organo-mineral fertilizer (F4) gave the best results, with 187.31 and 205.30 g per plant, respectively, but without differing statistically from that obtained with mineral fertilizer. Regardless of the source of fertilizer used, the grain yield was lower than that reported in the literature, 913.39 g per plant of papaya fertilized with cattle manure (Fernandes et al., 2010). However, considering the conditions under which the experiment was conducted (rainfed), the yield of 205.30 g/plant (F4) corroborated the data of Souza et al. (2007), who, under similar climatic conditions, obtained a yield of 266.04 g, on this occasion, with mineral fertilization.

There is also a relationship between grain yield and the number of racemes produced per plant, showing that the highest averages for both variables were obtained using the sources F4 > F2 > F3 > F1.

Regardless of the source of fertilizer used, the average number of racemes harvested was higher than that referenced for the cultivar "BRS 149 Nordestina" which, according to Embrapa Cotton (2002), produces an average of 5.2 racemes under rainfed conditions. In the absence of fertilization (F1), the number of racemes was considered medium (5 to 7) and with the use of sources F2, F3 and F4 it was high (> 7), according to the classification proposed by Savy Filho et al. (1999).

In general, the highest fruit yields per raceme were obtained from plants fertilized with the F4 (20), F2 (17) and F3 (12) sources. However, these yields were considered low when compared to the 35 and 37 fruits per raceme according to Embrapa Cotton (2002) and Carvalho (2005), respectively. These variations are certainly due to the soil and climate conditions, the cultural management applied, as well as the number of raceme orders considered in each study for calculating the characteristic in question.

The averages for percentage grain yield (PGY) did not differ statistically between fertilizer sources. These results show that under favorable conditions, regardless of the source of fertilizer used, the RPG will remain practically the same. It is worth noting that the order of the raceme was not taken into account when determining this variable, so the productivity and percentage grain yield data represent a general average, depending on the fertilizer source. This information is important since Banzatto et al. (1976) state that the percentage of unhatched seeds increases from the first raceme to the subsequent ones.

In general, the grain yield values found in this work are in line with those cited by Duke (1983), which range from 65% to 75%, depending on the degree of ripeness of the fruit, the cultivar and the environmental conditions.

References

Banzatto, N. V.; Zink, E.; Savy Filho, A. The contribution of primary, secondary and tertiary racemes in castor bean seed production. Semente, v. 2, n. 2, p. 32-34, 1976.

Carvalho, B. C. L. Manual do cultivo da mamona. Salvador: EBDA, 2005. 65p. il

Chaves, L. H. G.; Fernandes, J. D.; Dantas, J. P.; Monteiro Filho, A. F.; Silva, J. R. P.; Mesquita, E. F. Effect of different fertilizer sources on the production of castor bean cultivar BRS Nordestina. Cadernos de Agroecologia, v. 6, n. 2, p. 1-5, 2011.

Duke, J. A. Ricinus communis L. Handbook of Energy Crops. Purdue, 1983. Available at:

http://www.hort.purdue.edu/newcrop/duke energy/Ricinus communis.html. Accessed on: June 28, 2011

Embrapa Cotton. BRS-149 Nordestina. Campina Grande: Embrapa Cotton. 2002. 1 folder.

Embrapa - Brazilian Agricultural Research Corporation. National Soil Research Center. Manual of soil analysis methods. Rio de Janeiro: EMBRAPA. 212p. 1997.

Fernandes, J. D. ; Chaves, L. H. G.; Dantas, J. P.; Silva, J. R. P. Production components of brs nordestina grown with different fertilizer sources. In: IV Congresso Brasileiro de Mamona e I

Simpòsio Internacional de Oleaginos Energéticas, 2010, Joao Pessoa. Social Inclusion and Energy. Campina Grande: Embrapa Cotton, 2010. p. 486-491.

Ferreira, D.F. Sisvar: a computer statistical analysis system. Ci. Agrot., v. 35, p.1039-1042, 2011.

Savy Filho, A.; Banzatto, N. V.; Veiga, R. F. A.; Chiavegato, E. J.; Camargo, C. E. O.; Campo-Dall'orto F. A.; Godoy, I. J.; Fazuoli, L. C.; Carbonel, S. A. M.; Siqueira, W. J. Descritores minimos para o registro institucional de cultivares: Mamona. Campinas: IAC, 1999. 7p. (Documentos IAC, 61).

Souza, A. S.; Tàvora, F. J. A. F.; Pitombeira, J. B.; Bezerra, F. M. L. Planting times and irrigation management for papaya. I - production components. Revista Ciência Agronômica, v. 38, n. 4. p. 414-421, 2007.

CHAPTER 5

PHYTOREMEDIATION OF CONTAMINATED SOILS USING CASTOR BEAN

Phytoextraction is a promising phytoremediation technique that seeks to extract heavy metals from contaminated soils. The advantages of phytoremediation include minimal degradation and destabilization of the area, low environmental impact and favourable aesthetics (Nedelkoska and Doran, 2000). In addition, it is a clean, low-cost alternative (Tandy et al., 2004), provides leachate containment, maintains and improves the soil's physical structure, fertility and biodiversity, and absorbs metals from the soil that are expensive to extract using other technologies (Khan et al., 2000).

Among the many plants that have so far been identified as hyperaccumulators are some belonging to the Euphorbiaceae family (Garbisu and Alkorta, 2001). The castor bean tree (Ricinus communis L.), which belongs to this family, is a plant with a high degree of regeneration and biomass production, which can be cultivated in unfavorable conditions. However, the results of research into the extraction of heavy metals by this plant are still incipient and preliminary.

Khan et al. (1998) showed that castor bean is capable of removing zinc, cadmium and nickel from contaminated soils; Shi & Cai (2009) also refer to castor bean as a moderately tolerant plant to cadmium toxicity; according to Romeiro et al. (2006) and Liu et al. (2008), castor bean was considered a hyperaccumulator of lead, showing tolerance to this element at low concentrations.

The elements copper and zinc are essential for plant growth when they are at adequate levels. However, high concentrations of these metals in soils, in their available forms, can be harmful to plant development. For this reason, experiments were carried out using castor beans for soil phytoremediation.

CASE STUDY 9

With a view to castor bean plants being used in the phytoremediation of soils contaminated by copper and zinc, the aim of this study was to evaluate the accumulation and translocation of these elements in parts of the BRS Paraguaçu castor bean plant. To this end, two experiments were carried out in a greenhouse belonging to the Department of Agricultural Engineering at the Federal University of Campina Grande, using plastic pots with a capacity of 25 L, which were filled with 22 kg of sandy loam soil material (Quartzarenic Neosol), collected from the topsoil, whose characteristics were determined according to Embrapa (1997): pH (H_2O) = 5.8 ; Ca = 2.47 cmolc kg^{-1} ; Mg = 2.57 cmolc kg^{-1} ; Na = 0.52 cmolc kg^{-1} ; K = 0.19 cmolc kg^{-1} ; H = 0.76 cmolc kg^{-1} ; Al = 0.0 cmolc kg^{-1} ; organic matter = 3.8 g kg^{-1} ; P = 5.9 mg kg^{-1} ; Zn = 4.17 mg dm^{-3} ; Cu = 6.23 mg dm^{-3} . The experiments were set up in a completely randomized design with three replications, one with five levels of copper (0; 25; 50; 75 and 100 mg kg^{-1}) and the other with five levels of zinc (0; 50; 100; 150 and 200 mg kg^{-1}

), making up a total of fifteen experimental units for each, using copper chloride and zinc chloride as sources of the elements, respectively, which were applied in a mixture to the soil before planting. The soil in each experimental unit received an NPK fertilizer corresponding to the plants' needs. Each experimental unit received three seeds of castor bean cultivar BRS Paraguaçu and, after thinning, one plant remained per unit. Throughout the experimental period (120 days) the soil was kept at 80% of the available water. In order to assess the nutritional diagnosis of the plant in relation to the absorption of copper and zinc throughout the experimental period, at 30; 60; 75 and 105 days after sowing (DAS) one leaf per plant was collected - 4ª . leaf from the apex. At the end of the experiment, the plants were harvested from each experimental unit and the plant material was separated into roots, stems and leaves. All this material was prepared and subjected to nitroperchloric digestion for the determination of copper and zinc levels by atomic absorption spectrophotometry. At the end of the experiment, the plants were harvested and the plant material was separated into roots, stems and leaves, which were then subjected to the procedures described above. The accumulation (A) of copper and zinc in the leaves, stems and roots of the plants (μg /plant part) was calculated based on the concentration of the element and the dry mass in each plant part. Using the total amount accumulated, the percentage of metal accumulation for each plant part was calculated. According to Abichequer & Bohnen (1998), the translocation index (TI) of the elements was calculated using the equation TI= (Amount accumulated in the aerial part / Amount accumulated in the plant). The data was analyzed using analysis of variance and regression analysis for significant data, using the Assistat software version 7.5 beta (2008).

The phytomass of the leaves, stem and roots of the papaya tree was influenced by the doses of copper, with the treatment corresponding to 50 mg kg^{-1} providing the highest values (Table 1). From this treatment onwards, the phytomass decreased, signaling a probable detrimental effect of copper on the growth of the papaya tree.

Table 1. Significance of Mean Squares and weight of phytomass of leaves, stem and roots of castor bean plants as a function of copper and zinc treatments

Treatments mg kg^{-1}	Phytomass, g $plant^{-1}$ Leaf	Stem	Root
		Copper	
	*	*	*
0	42.21 ab	37.93 ab	32.32 ab
25	36.67 ab	33.36 ab	28.13 ab
50	51,52 a	43,87 a	37,12 a
75	38.58 ab	25.72 ab	26.60 ab
100	26,83 b	20,30 b	18,38 b
		Zinc	
	ns	ns	ns
0	42,21 a	37,93 a	32,32 a
50	33,46 a	31,31 a	31,10 a
100	45,02 a	42,91 a	42,24 a
150	41,38 a	40,75 a	40,59 a

200	42,78 a	29,48 a	32,81 a

* significant at 5% probability; ns not significant according to the F test. Source: Chaves et al. (2010)

The doses of zinc did not significantly influence the phytomass production of the castor bean plant; however, there was a tendency for the phytomass to increase up to the dose of 100 mg kg^{-1} of the element. Zeitouni et al. (2007), evaluating the effects of zinc and copper on castor bean and sunflower, observed a decrease in plant dry matter as the doses of the metals applied to the soil increased, showing that the reference values for soil quality for zinc and copper correspond to 60 and 35 mg kg^{-1}, respectively (Cetesb, 2007).

The copper levels found in the 4ª . leaves of the plants throughout the experimental period show that at the beginning of the crop cycle the plants absorbed greater amounts of the element, in proportion to their development, and as the plants grew there was a general tendency for copper absorption and/or translocation to the leaves to decrease (Figure 1).

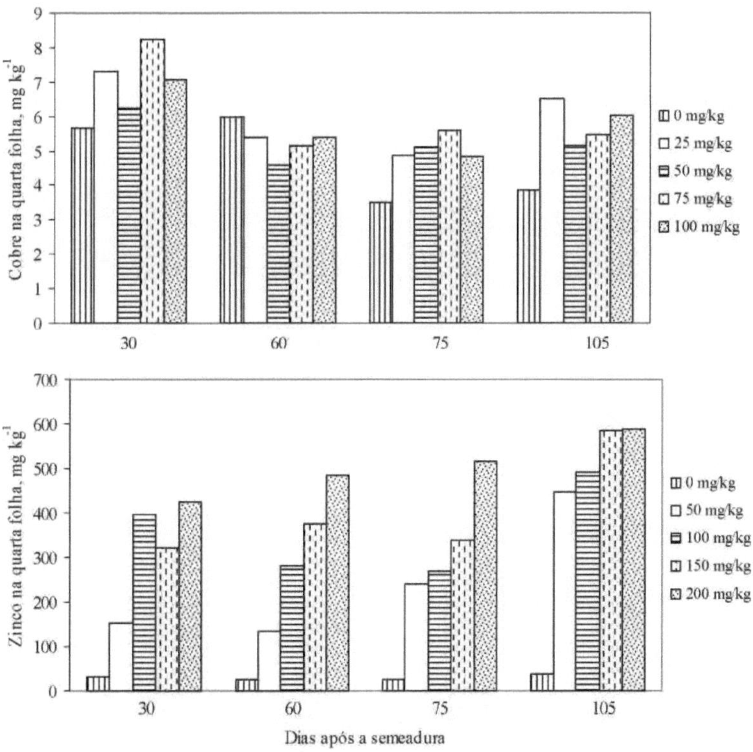

Figure 1: Copper and zinc content in the 4th leaf of castor bean plants subjected to increasing metal treatments over the experimental period. Source: Chaves et al. (2010)

With regard to the zinc levels found in the 4ª . leaves of the plants, it can be seen that they generally increased throughout the experimental period, showing that the plants continue to absorb and/or translocate the element increasingly, even after flowering (around 75 DAS).

The copper levels in the leaves, stems and roots of the castor bean plants, collected at the end of the experiment, were influenced by the copper treatments (Table 2), which increased quadratically as a function of the treatments (Figure 2 a, b) with the exception that with the highest dose of copper, there was a decrease in the content of the element in the stem. Comparing the levels of copper in the leaves and roots of the plants that received the highest dose in relation to the control, it was observed that there was an increase in the levels of the element of 265% and 4379%, respectively. In the stem, comparing the highest content observed (treatment 75 mg kg^{-1}) with the control, there was an increase of 106%.

Table 2. Significance of the Mean Squares and the accumulated copper and zinc in the leaves, stem and roots of castor bean plants, as a function of the treatments.

	Accumulated, µg plant $part^{-1}$		
Treatments mg kg^{-1}	Leaf	Stem	Root
		Copper	
	ns	**	**
0	130,35 a	46,57 b	205,49 b
25	159,00 a	62.47 ab	1101,05 b
50	318,71 a	106,21 a	2423.09 ab
75	240,37 a	63.16 ab	4820,81 a
100	302,17 a	25,37 b	5063,69 a
		Zinc	
	**	**	**
0	2479,20 d	949,51 c	2079,31 c
50	13279.38 cd	11585,08 b	21899,56 c
100	24247.41 bc	22298,76 a	80923,03 b
150	27315.42 ab	23578,08 a	128211,50 a
200	38389,96 a	20519,59 a	111994.10 ab

* *significant at 1% probability; ns not significant, with F test.

Source: Chaves et al. (2010)

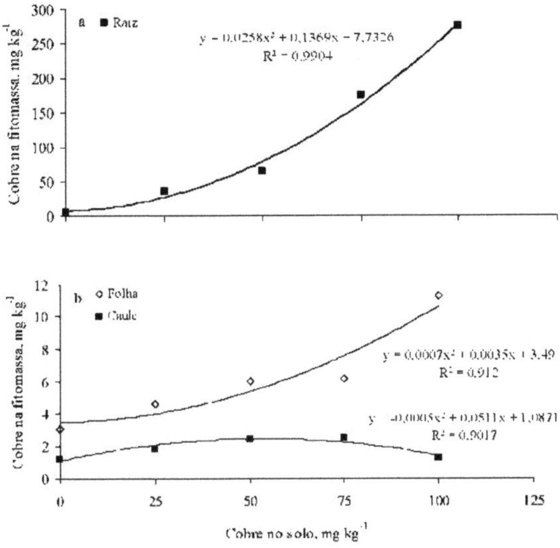

Figura 2 - Copper content in the root (a), leaves and stem (b) of castor bean as a function of the doses of copper applied to the soil. Source: Chaves et al. (2010)

With the exception of the control plant, the levels found in the leaves and stems were in the range of 100 to 400 mg kg^{-1} or higher, even though there was no decrease in productivity (Table 2). In the aerial part as a whole (sum of the zinc contents present in the leaf, ranging from 57.32 to 898.25 mg kg^{-1} and in the stem from 25.00 to 697.17 mg kg^{-1}), the element content varied on average from 83 to 1595 mg kg^{-1} corroborating Zeitouni et al. (2007).

The treatments had a significant effect on the zinc levels found in the leaves, stem and roots (Table 2). According to the regression analysis, the zinc levels in the leaves increased linearly, while in the stems and roots they increased quadratically (Figure 3), with an increase of up to 14.67 x 100% in the leaves, 26.89 x 100% in the stems and 53.71 x 100% in the roots, when comparing the zinc levels in the respective parts of the plants that received the highest treatment in relation to the control. Also for this element, the highest levels were observed in the roots of the plants, corroborating Liu et al. (2008).

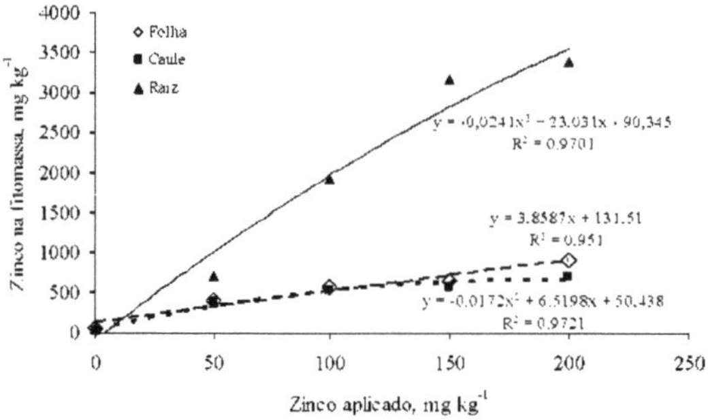

Figura 3 - Zinc content in castor bean leaves, stems and roots as a function of zinc doses applied to the soil. Source: Chaves et al. (2010)

The translocation index measures the plant's efficiency in transporting an element from the root to the aerial part. This index decreased as the doses of copper and zinc increased, ranging from 0.48 to 0.06 and from 0.61 to 0.28, respectively. This index being higher than 1 (reference value) indicates the plant's efficient ability to transport metals from the roots to the aerial part, confirming the plant's hyperaccumulation characteristic (Mahmud et al., 2008, Melo et al., 2009). Figure 4 shows the percentage values of the translocation index of copper and zinc in plants subjected to increasing levels of these elements. In the case of copper, it can be seen that in the plants subjected to the two highest levels of the element, translocation from the roots to the aerial part was lower than in the other levels. It is worth noting that, in general, from the treatment corresponding to 50 mg kg^{-1} , as the levels of copper applied to the soil increased, the values accumulated in the aerial parts of the plants decreased (Table 2), which is justified by the decrease in the translocation rate of the element.

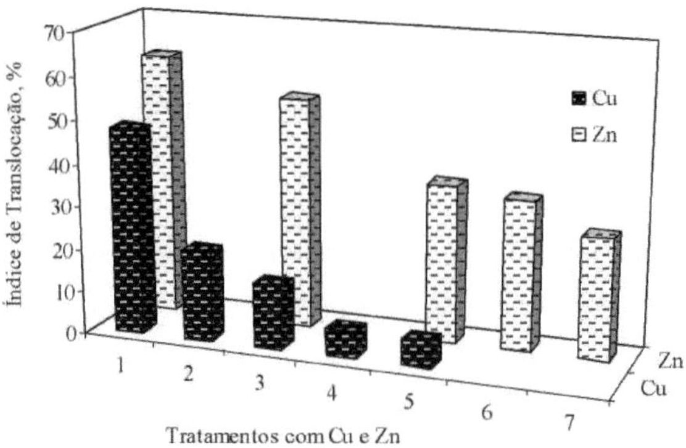

Figura 4 - translocation index of copper and zinc in castor beans subjected to increasing levels of these elements applied to the soil (1= control; 2 = 25 mg Cu kg^{-1} ; 3 = 50 mg Cu, Zn kg^{-1} ; 4= 75 mg Cu kg^{-1} ; 5 = 100 mg Cu, Zn kg^{-1} ; 6 = 150 mg Zn kg^{-1} ; 7 = 200 mg Zn kg^{-1}). Source: Chaves et al. (2010)

With regard to zinc, the variation in translocation indices as a function of treatment was gradual and less marked than that observed for copper (Figure 4).

Comparing the translocation indices of the two elements (Figure 4), those corresponding to zinc were higher than those corresponding to copper, especially for equal levels of the elements added to the soil (50 and 100 mg kg^{-1} of Cu and Zn).

The copper removal values in the 0; 25; 50; 75 and 100 mg kg^{-1} treatments were 0.28; 0.24; 0.26; 0.33 and 0.23%, respectively. For zinc, in the treatments 0; 50; 100; 150 and 200 mg kg^{-1} , the removal values were 6.0; 4.3; 5.8; 5.4 and 3.9%, respectively. Kabata-Pendias (1995) considers a plant to be a zinc accumulator when it removes around 1% of the total zinc present in the soil. In this experiment, castor bean had higher zinc removal values than those described above, so it could be considered an accumulator, corroborating Zeitouni (2003). Based on the value of 1% copper removal as a reference, castor bean is not considered to be an accumulator of this element, since the values of copper removal by the plant were below the reference value.

Based on the results, it can be concluded that the accumulation of copper and zinc in castor bean plants followed the following order: root > leaf > stem. The translocation of zinc was greater than that of copper in castor bean plants. The castor bean in this study does not fit the definition of a hyperaccumulator plant, but it may be a zinc accumulator.

References

Abichequer, A. D.; Bohnen, H. Efficiency of absorption, translocation and utilization of phosphorus by wheat varieties. Revista Brasileira de Ciência do Solo, v.22, p.21-26, 1998.

Cetesb. Environmental Sanitation Technology Company. List of contaminated areas in the state of Sao Paulo. Available at: http:/www.cetesb.sp.gov.br. Accessed on January 29, 2009.

Chaves, L. H. G.; Mesquita, E. F.; Araujo, D. L.; França, C. P. Accumulation and distribution of copper and zinc in papaya cultivar BRS Paraguaçu and plant growth. Engenharia Ambiental, v. 7, n. 3, p. 263-277, 2010.

Embrapa. National Soil Research Center. Manual of soil analysis methods. 2.ed. Rio de Janeiro: Embrapa, CNPS, 1997. 212p.

Kabata-Pendias, A. Agricultural problems related to excessive trace metal contents of soils. In: Salomons, W.; Forstner, U.; Mader, P. Heavy metals: problems and solutions. Berlin: Springer - Verlag, 1995. p.3-18.

Liu, X.; Gao, Y.; Khan, S.; Duan, G.; Chen, A.; Ling, L.; Zhao, L.; Liu, Z.; Wu, X. Accumulation of Pb, Cu and Zn in native plants growing on contaminated sites and their potential accumulation capacity in Heqing, Yunnam. Journal of Environmental Science, v. 20, n.12, p. 1469-1474, 2008.

Mahmud, R.; Inoue, N.; Kasajima, S.Y.; Shaheen, R. Assessment of potential indigenous plant species for the phytoremediation of arsenic contaminated areas of Bangladesh. International Journal Phytoremediation, v.10, p.119-132, 2008.

Melo, E. E. C.; Costa, E.T.; Guilherme, L.R.; Faquim, V.; Nascimento, C.W. Accumulation of arsenic and nutrients by castor bean plants grown on an As- enriched nutrient solution. Journal of Hazardous Materials, v.168, p. 479-483, 2009.

Zeitouni, C. F. Efficiency of plant species as phytoextractors of cadmium, lead, copper, nickel and zinc from a dystrophic Yellow Red Latosol. 2003. 99 f. Dissertation (Master's Degree) - Agronomic Institute, Campinas, 2003.

Zeitouni, C. F.; Berton, R. S.; Abreu, C. A. Phytoextraction of cadmium and zinc from a red-yellow latosol contaminated with heavy metals. Bragantia, v.66, n.4, p. 649-657, 2007.

CHAPTER 6

FINAL CONSIDERATIONS

This book presented several studies showing the agronomic viability of castor bean cultivation from different perspectives. The results presented here indicate that Brazil is in a position to expand the cultivation of castor bean as a raw material in order to meet the supply of both biodiesel and lubricants, especially in Brazil's semi-arid regions, where the crop is well adapted.

The studies on the fertilization of castor beans showed that the highest doses of nitrogen, phosphorus and potassium provided the highest dry phytomass values, with the crop being more responsive to doses of nitrogen than to doses of phosphorus and potassium. Nitrogen was the nutrient that promoted the highest yield of castor bean, while doses of potassium, copper and zinc provided the highest production of seed weight.

With regard to nitrogen fertilization versus water consumption, the results showed that there was an increase in phytomass, water consumption and water use efficiency as the doses of nitrogen applied to the soil increased, indicating high efficiency in transforming the water consumed into dry matter for the papaya tree.

The results of the studies on the organic fertilization of castor bean trees showed that the castor oil obtained can be classified commercially as industrial type 1, as it is below the maximum permitted limit of 0.5% humidity, and also shows favorable results for the vegetative growth of the plants.

Phytoremediation using castor bean plants showed positive results in terms of copper and zinc accumulation, with zinc translocation being greater than that of copper, characterizing castor bean as a zinc-accumulating plant.

This shows that castor bean has favorable agronomic characteristics for cultivation in Brazil's semi-arid regions. It is therefore necessary to adopt public policies in order to increase castor bean production and its commercialization, so as to guarantee family farmers an income that makes it viable to use their land to produce a crop that does not serve as food for the population.

In this sense, the studies presented in this book help to ensure that castor beans can be grown under the right fertilization and irrigation conditions for good production, giving family farmers an alternative way of planting their land.

BIBLIOGRAPHY CONSULTED

Azevedo, D. M.; Lima, E. F.; Batista, F. A. S.; Beltrao, N. E. M.; Soares, J. J.; Vieira, R. M.; Moreira, J. A. M. Recomendações técnicas para o cultivo da mamoneira (Ricinus communis L.) no nordeste do Brasil. Campina Grande: Embrapa-CNPA, 1997. 52 p. (Technical Circular, 25).

Azevedo, D. M. P.; Lima, E. F. (ed.). The Castor Bean Agribusiness in Brazil. Campina Grande: Embrapa Algodao, 2001. 350 p.

Garbisu, C.; Alkorta, I. Phytoextraction: a cost effective plant-based technology for the removal of metals from the environment. Bioresource Technology, v. 77, n. 3, p. 229-236, 2001.

Khan, A.G.; Chaudhry, T.M.; Hayes, W.J.; Khoo, C.H.; Hill, L.; Fernandez, R.; Gallardo, P. Physical, chemical and biological characteristics of a steelworks waste site at Port Kembla, NSW, Australia. Water, Air Soil Pollution, v. 104, n. 3-4, p.389-402, 1998.

Khan, A. G.; Kuek, C.; Chaudhry. T.M.; Khoo, C.S.; Hayes, W. J. Role of plants, mycorrhizae and phytochelators in heavy metal contaminated land remediation. Chemosphere, v.41, n.1-1, p.197-207, 2000.

Lange, A.; Martines, A. M.; Silva, M. A. C.; Sobreano, M. C. M.; Cabral, C. P.; Malavolta, E. Effect of micronutrient deficiency on the nutritional status of the papaya cultivar Iris. Pesquisa Agropecuâria Brasileira, v. 40, p. 61-67, 2005.

Liu, X.; Gao, Y.; Khan, S.; Duan, G.; Chen, A.; Ling, L.; Zhao, L.; Liu, Z.; Wu, X. Accumulation of Pb, Cu and Zn in native plants growing on contaminated sites and their potential accumulation capacity in Heqing, Yunnam. Journal of Environmental Science, v. 20, n.12, p. 1469-1474, 2008.

Malavolta, E.; Vitti, G. C.; Oliveira, S. A. Evaluation of plant nutritional status: principles and applications. 2.ed. Piracicaba: POTAFOS, 1997. 201p.

Nakagawa, J.; Levorato, E.; Boaretto, A. E. Effects of increasing doses of thermophosphate in the presence and absence of micronutrients on two cultivars of papaya *(Ricinus communis* L.). Cientifica, Sao Paulo, v. 14, p. 55-64, 1986.

Nedelkoska, T. V.; Doran, P. M. Characteristics of heavy metal uptake by plant species with potential for phytoremediation and phytomining. Minerals Engineering, v.13, p. 549- 561, 2000.

Oliveira, F.A.; Oliveira Filho, A. F.; Medeiros, J. F.; Almeida Junior, A. B.; Linhares, P. C. F. Initial development of papaya under different sources and doses of organic matter. Caatinga, v. 22, n. 1, p. 206-211, 2009.

Paulo, E. M.; Bataglia, O. C.; Kasai, F. S.; Cavichioli, J. C. Boron deficiency in castor bean. Bragantia, Campinas, v. 48, p. 241-247, 1989.

Romeiro, S.; Lagôa, A.M.M.A.; Furlani, P.R.; Abreu, C.A.; Abreu, M.F.; Erismann, N.M. Lead and tolerance of Ricinus communis L. Brazilian Journal of Plant Physiology, v. 18, n.4, p. 483-489, 2006.

Santos, R.F.; Barros, M.A.L.; Marques, F.M.; Firmino, P.T. Requiao, L.E; G. Anàlise Econòmica In: Azevedo, D. M. P.; Lima, E. F. (eds). The castor bean agribusiness in Brazil. Brasilia: EMBRAPA technological communication service, 2001. 350p.

Savy Filho, A.; Banzatto, N. V.; Veiga, R. F. A.; Chiavegato, E. J.; Camargo, C. E. O.; Campo-Dall'Orto F. A.; Godoy, I. J.; Fazuoli, L. C.; Carbonel, S. A. M.; Siqueira, W. J. Descritores minimos para o registro institucional de cultivares: Mamona. Campinas: IAC, 1999. 7p. (Documentos IAC, 61).

Severino, L. S.; Ferreira, G. B.; Moraes, C. R. A.; Gondim, T. M. S.; Freire, W. S. A.; Castro, D. A.; Cardoso, G. D.; Beltrao, N. E. M. Growth and productivity of papaya fertilized with macronutrients and micronutrients. Pesquisa Agropecuària Brasileira, v. 41, n. 4, p. 563-568, 2006.

Shi, G.; CaI, Q. Cadmium tolerance and accumulation in eight potential energy crops. Biotechnology Advances, v. 27, p.555-561, 2009.

Silva, M. F. O.; Prates, E. M.; Sirqueira, F. J. S.; Soares, N. S. Analysis of the Brazilian castor bean supply from 1961 to 2012. Estudo & Debate, v. 22, n. 2, p. 195-213, 2015.

Souza, E. A.; Natale, W. Effect of boron and zinc on papaya cultivation. Cientifica, Sao Paulo, v. 2, p. 327-333, 1997.

Tandy, S.; Bossart, K.; Mueller, R.; Ritschel, J.; Hauser, L.; Schulin, R.; Nowack, B. Extraction of heavy metals from soils using biodegradable chelating agents. Environmental Science Technology, v. 38, n.3, p. 937- 944, 2004.

Tàvora, F. J. A. F. A. The castor bean crop. Fortaleza: EPACE, 1982. 111p.

MIX
Papier aus verantwortungsvollen Quellen
Paper from responsible sources
FSC® C105338

Printed by Books on Demand GmbH, Norderstedt / Germany